Manish Kumar Singh
Priyanka Singh
Dr. Shrikant Chitale

Intercropping Under Rice-Based Cropping System

An Experimental Study on Productivity and Profitability

Anchor Academic Publishing

Singh, Manish Kumar, Singh, Priyanka, Dr. Chitale, Shirkant: Intercropping Under Rice-Based Cropping System: An Experimental Study on Productivity and Profitability. Hamburg, Anchor Academic Publishing 2013

Buch-ISBN: 978-3-95489-122-1
PDF-eBook-ISBN: 978-3-95489-622-6
Druck/Herstellung: Anchor Academic Publishing, Hamburg, 2013

Bibliografische Information der Deutschen Nationalbibliothek:
Die Deutsche Nationalbibliothek verzeichnet diese Publikation in der Deutschen Nationalbibliografie; detaillierte bibliografische Daten sind im Internet über http://dnb.d-nb.de abrufbar.

Bibliographical Information of the German National Library:
The German National Library lists this publication in the German National Bibliography. Detailed bibliographic data can be found at: http://dnb.d-nb.de

C O N T E N TS AT A GLANCE

CHAPTER	PARTICULARS	PAGE No.
I.	INTRODUCTION	1 - 11
II.	REVIEW OF LITERATURE	12 - 32
III.	MATERIALS AND METHODS	33 - 52
IV.	RESULTS AND DISCUSSION	53 - 83
V.	SUMMARY, CONCLUSIONS AND SUGGESTIONS FOR FUTURE RESEARCH WORK	84 - 94
	ABSTRACT	95 - 96
	REFERENCES	97 - 107
	APPENDICES	108 - 118

DETAILED C O N T E N TS

CHAPTER	PARTICULARS	PAGE No.
I.	**INTRODUCTION**	**1 - 11**
1.1	**Intercropping**	**1-3**
1.2	**Objectives of intercropping**	**3**
1.3	**Types of intercropping**	**3**
1.4	**Intercropping concepts**	**3**
1.4.1	Spatial arrangement	4
1.4.2	Plant density	5
1.4.3	Maturity dates	5
1.4.4	Plant architecture	5
1.5	**Advantages of intercropping**	**5**
1.6	**Disadvantages of intercropping**	**7**
1.7	**Intercropping : global scenario**	**7**
1.8	**Intercropping in India**	**8**
II.	**REVIEW OF LITERATURE**	**12– 32**
2.1	**Effect of cropping systems on**	**13**
2.1.1	Growth and yield of rice	13
2.1.2	Total productivity of system including field crops	13
2.1.3	Total productivity of system including vegetables	15
2.1.4	Total productivity of system with intercrop	18
2.2	Effect of cropping systems on soil fertility status	19
2.3	Effect of cropping systems on weed dynamics	21
2.4	Water use efficiency	23
2.5	Economic viability	24

CHAPTER	PARTICULARS	PAGE No.
2.6	Employment generation, production and land utilization efficiency	28
2.7	Energetics	31
III.	**MATERIALS AND METHODS**	**33 - 52**
3.1	Geographical Situation	33
3.2	Climatic condition	33
3.3	Weather condition during crop growth	33
3.4	Cropping history of the experimental field	35
3.5	Physico-chemical characteristics of soil	35
3.6	Experimental details	36
3.7	**Test crops**	**39**
3.7.1	**Rice (*Oryza sativa*)**	**39**
3.7.2	**Wheat (*Triticum aestivum* L.)**	**39**
3.7.3	Castor (*Ricinus communis*)	39
3.7.4	Lentil (*Lens culinaris* MediK.)	39
3.7.5	Mustard (*Brassica juncea*)	39
3.7.6	Sunflower (*Helianthus annuus* L.)	41
3.7.7	Fenugreek (*Trigonella foenumgraceum*)	41
3.7.8	Onion (*Allium cepa* L.)	41
3.7.9	Coriander (*Coriandrum sativum* L.)	41
3.8	Experimental details and cultural operations	41
3.9	Seed treatment	43
3.10	Transplanting of rice	43
3.11	Cultural Schedule	43
3.12	Harvesting and threshing	43
3.13	**Studies on crops**	**46**
3.13.1	Pre-harvest observation	46
3.13.1.1	Plant population (No. m^{-2})	46
3.13.1.2	Plant height (cm)	46
3.13.1.3	Dry matter accumulation (g $plant^{-1}$)	46

CHAPTER	PARTICULARS	PAGE No.
3.13.1.4	Leaf area index (LAI)	46
3.14	Weed studies	46
3.14.1	Weed density	47
3.14.2	Dry weight of weeds	47
3.15	**Post harvest observations**	**47**
3.15.1	Yield components	47
3.15.1.1	Panicles plant^{-1}	47
3.15.1.2	Grains panicle^{-1}	47
3.15.1.3	Panicle length (cm)	47
3.15.1.4	Test weight (g)	48
3.15.2	**Biomass production**	**48**
3.15.3	**Grain yield (q ha^{-1})**	**48**
3.15.4	Straw yield (q ha^{-1})	48
3.15.5	Harvest index (%)	48
3.15.6	**Wheat equivalent yield (kg ha^{-1})**	**48**
3.16	**Chemical analysis**	**49**
3.16.1	Organic carbon content	49
3.16.2	Available nitrogen	49
3.16.3	Available phosphorus	49
3.16.4	Available potassium	49
3.17	**Economic analysis**	**49**
3.18	**System analysis**	**50**
3.18.1	Productivity efficiency (PE)	50
3.18.2	Economic efficiency	50
3.18.3	Relative productivity efficiency (RPE) and relative economic efficiency (REE)	50
3.18.4	Irrigation water use efficiency	51
3.18.5	Employment generation efficiency	51
3.19	**Energetics**	**51**

CHAPTER	PARTICULARS	PAGE No.
3.20	**Satistical analysis**	**52**
IV.	**RESULTS AND DISCUSSION**	**53 - 83**
4.1	**Studies in rice**	**53**
4.1.1	Plant population and plant height of rice	53
4.1.2	Dry matter accumulation of rice (g plant -1)	56
4.1.3	Leaf area index of rice	57
4.1.4	Yield attributing characters of rice (Number of panicles plant -1)	60
4.1.5	Grain and straw yield (q ha-1) and harvest index (%) of rice	61
4.1.6	Weed studies in rice	64
4.1.7	Available nutrient status at kharif harvest	64
4.1.8	Economics of rice	66
4.2	**Studies in rabi crops**	**67**
4.2.1	Grain yield in terms of wheat equivalent yield	68
4.2.2	Weed dynamics in rabi crops	69
4.2.3	Available nutrient status at rabi harvest	71
4.2.4	Economics of rabi crops	72
4.3	**Total productivity and system Analysis**	**74**
4.3.1	Total productivity in terms of WEY	74
4.3.2	Economics of the system	75
4.3.3	Production efficiency and economic efficiency	79
4.3.4	Relative productivity efficiency and relative economic efficiency	80
4.3.5	Employment generation efficiency	80
4.3.6	Irrigation water use efficiency and irrigation water requirement	82
4.3.7	Energetics	83
V.	**SUMMARY, CONCLUSIONS AND SUGGESTIONS FOR FUTURE RESEARCH WORK**	**84 - 94**
	ABSTRACT	**95 - 96**
	REFERENCES	**97 - 107**
	APPENDICES	**108 - 118**

LIST OF TABLES

TABLE No.	PARTICULARS	PAGE No.
3.1	Physico-chemical properties of the experimental site.	36
3.2	Treatment details of experiment.	38
3.3.	Details of original experiment.	38
3.4	Experimental details of fertilizer doses, sowing and harvesting dates.	42
3.5	Schedule of different cultural operation during *kharif* and *rabi* season.	44 - 45
4.1	Plant population of three rice varieties as affected by different cropping systems	54
4.2	Plant height (cm) of rice varieties as affected by different cropping systems	55
4.3	Dry matter accumulation plant $^{-1}$ of three rice verities as affected by different cropping systems.	56
4.4	Leaf area index (LAI) of three rice varieties as affected by different cropping system.	58
4.5	Yield attributing characters of three rice varieties as effected by different cropping systems.	60
4.6	Grain yield, straw yield, and harvest index of three rice varieties as affected by different cropping systems.	62
4.7	Total weed population and weed dry weight of weeds as affected by different cropping systems	63
4.8	Effect of different cropping systems on organic carbon and available NPK content in soil at the time of *kharif* harvest.	65
4.9	Cost of cultivation, net return and benefit: cost ratio of rice as affected by different cropping systems.	67
4.10	Grain yield and wheat equivalent yield of *rabi* crops as influenced by different cropping systems.	69

TABLE No.	PARTICULARS	PAGES No.
4.11	Total weed population and dry matter of *rabi* season weeds as affected by different cropping systems.	70
4.12	Available N, P and K and organic carbon (OC) content of soil after *rabi* harvest as affected by different cropping systems.	72
4.13	Cost of cultivation, net return and B:C ratio of *rabi* crops as influenced by different cropping systems.	73
4.14	Total Productivity (TP) of the system in terms of wheat equivalent yield (WEY), total cost of cultivation, total net return and benefit: cost ratio of different rice-based cropping systems.	76
4.15	Production efficiency, economic efficiency, relative productivity efficiency and relative economic efficiency as affected by different cropping systems.	77
4.16	Employment generation efficiency, no. of labour employed, irrigation water use efficiency and irrigation water requirement as affected by different cropping systems.	81
4.17	Total energy input, output, input: output ratio and energy use efficiency of different rice based cropping systems.	83

LIST OF FIGURES

FIGURE No.	PARTICULARS	PAGES IN BETWEEN
3.1	Weekly meteorological data during crop growth period (From July 01, 2009 to March 30, 2010)	34
3.2	The layout plan and other details of experiment	37
4.1	Leaf area index of three rice varieties as affected by different cropping systems.	59
4.2	Grain yield of rice, *rabi* crops and total productivity of the system in terms of wheat equivalent yield (WEY) as affected by different cropping systems	78
4.3	Relative productivity and relative economic efficiency of different cropping systems.	79

LIST OF PLATES

PLATE No.	PARTICULARS	PAGES IN BETWEEN
1	View of sunflower +lentil	40
2	View of wheat+ fenugreek	40

LIST OF APPENDICES

APPENDIX	PARTICULARS	PAGE No.
I	Weekly meteorological data during crop growth period (From July 01, 2009 to March 31, 2010).	108 – 109
II a	Cost of cultivation of rice (Rs. ha^{-1}) system of rice cucultivation.	110
II b	Cost of cultivation of wheat (Rs. ha^{-1})	111
II c	Cost of cultivation (Rs/ha) of wheat + lentil (1:1) skip/alternate row of 20 cm	112
II d	Cost of cultivation (Rs/ha) of wheat + lentil (1:1) skip/alternate row of 20 cm (Rs/ha) of castor + lentil (1:3)	113
II e	Cost of cultivation (Rs/ha) (Rs/ha) of onion +coriander (3:1)	114
II f	Cost of cultivation (Rs/ha) (Rs/ha) of wheat+ Fenugreek (1:1) skip/alternate row of 20 cm	115
II g	Cost of cultivation (Rs/ha) of mustard + lentil (1:2)	116
II h	Cost of Cultivation (Rs/ha) of sunflower + lentil (1:3)	117
II i	Prices of test crops	118

PREFACE

Intercropping or multiple cropping serves as an excellent strategy for intensifying land use and increasing income and production per unit area Therfore, this book emphasizes on the productivity and profitability of intercropping in rabi cereal, legume, oilseed and spices under rice (Oryza sativa L.) based cropping system. The major objectives were to assess the production potential and economic viability of different rabi intercropping under rice based cropping systems and to identify the suitable/ remunerative rice based cropping systems with vegetables and oilseeds intercrops.

"Education plays vital role in personal and social development and teacher plays a fundamental role in imparting education. Teachers have crucial role in shaping young people not only to face the future with confidence but also to build up it with aim and responsibility. There is no substitute for teacher pupil relationship". I take this golden opportunity to express my heartful, humble and deepest sense of gratitude to those who helped me to complete my research possible. These words are small acknowledgement but never fully recompensed for their great help and co-operation.

It is my privilege to study and conduct my research under ***Dr. Shrikant Chitale****, Scientist, Department of Agronomy, College of Agriculture, Raipur (C.G.), Chairman of my advisory committee, who provided me the research insight, illuminating and meticulous guidance, calm endurance, continuous and unfailing encouragement, scholarly suggestions, unique supervision, constructive criticisms, sympathetic attitude, plausible appreciation and sustained support during the entire course of investigation and preparation of manuscript. His scientific approach and generosity without any reservation have my privileges to work under his supervision, knowledge and enthusiastic interest, which he provided me throughout my post graduation and research investigation despite his busy schedule of work.*

"There is a place in the heart where thoughts become wishes and wishes become dreams." I express my sincere gratitude to my beloved father Shri S. N. Singh and mother Smt. Ranju Singh, who bore the weight of sacrifice with patience, whose selfless love, affection, sacrifices and blessing made my path easier and helped me to make my dreams come true. Their blessings have always been the most vital source of inspiration and motivation in my life. My most cordial thanks go to my younger sisters Priyanka and Namrata who inspired me constantly and moulded me into the present position.

"Performance = Individual attributes x *Work efforts* x *Friends support."There are many friends and well wishers who helped me in various ways towards the completion of this book and they deserve my sincere thanks. Last but not least, I would like to convey my cordial thanks to all the teachers and well wishers from my schooling days onwards who have directly and indirectly helped me to reach upto this level in my life. How can I express my thanks to "****God****" because there is no any word to express it. So, my lord, please realize and accept my feelings.*

Dedicated

To

My

Parents

INTRODUCTION

CHAPTER-I

INTRODUCTION

1.1 INTERCROPPING

In the era of shrinking resource base of land, water and energy, resource-use efficiency is an important aspect for considering the suitability of a cropping system (Yadav, 2002). Diversification and intensification of rice-based systems to increase productivity per unit resource is very pertinent. The diversification of cropping system is necessary to get higher yield and return, to maintain soil health, sustain environment and meet daily requirement of human and animals (Samui *et al.*, 2004).

Any modification to the existing system with a tendency to decline the productivity of rice crop will neither be sustainable nor acceptable to the farming community. Likewise, the importance of highly intensive crop sequence is also well recognized to meet the growing demand of ever-increasing population. An intensive cropping which must not only be highly productive and profitable but also be stable over the time and maintains soil fertility in present conditions (Ghosh, 1987). An intensification of cropping sequence is essentially depending on the need of the area. Oilseeds and pulses including vegetables are receiving more attention owing to higher prices due to increased demand. Inclusion of these crops in a sequence changes the economics of the cropping sequence. Moreover, diversification with inter-cropping can give higher yields than when grown as sole crops (Mandal *et al.*1986). Thus, selecting compatible combination of crops is necessary for the maximum utilization of growth resources, *viz.* solar energy, and water per unit area per unit time that will also keep the soil in a better physical condition as well as improvement in yield. Hence, choice of the component crops needs to be suitably maneuvered to harvest the synergism among them towards efficient utilization of resource base and to increase overall productivity (Anderson, 2005).

Rice farmers are mostly involved in monoculture practices. This deprives the land for growing other food crops. Hence, a better alternative of mono/sole cropping is required to overcome this shortcoming. Therefore, a shift from mono cropping to inter/multiple cropping as an excellent strategy for intensifying land use and increasing income and production per unit area and time is appreciated. Intercropping has been practiced all over the world from times not even known. Intercropping was also practiced in ancient Greece about 300 B.C. Theophrastus notes that wheat, barley, and certain pulses could be planted at various times during the growing season often integrated with vines and olives, indicating knowledge of the use of intercropping (Papanastasis et al., 2004). Intercropping can be defined as

Intercropping stands for growing more than one crop in the same field at the same time.

The crops are referred as

Base Crop: is the one which is plated as its optimum sole crop population in an intercropping situation and second crop is planted in between rows of base crop for obtaining bonus yield from intercrop without affecting base crop yield.

Component Crop: is used to refer to either individual crops making up the intercropping situation. Intercrop yield is the yield of a component crop when grown in intercropping and expressed over the total intercropped area. (i.e. area occupied by both the crops). A simple addition of both the intercrop yields a combined intercrop yield.

The component crops are selected on the following basis :

1. The time of peak nutrient demand should not overleap.
2. The competition for light should be minimum.
3. The component crops should be compatible with each other.
4. The maturity time of component crops should differ by at least one month.

1.2 OBJECTIVES OF INTERCROPPING

The objectives of Intercropping Systems are:

1. Indemnity against total crop failure.
2. Increase in crop production.
3. Optimum use of resources.

1.3 TYPES OF INTERCROPPING

Based on the per cent of plat population used in intercropping system, it is of two types:

- Additive series and
- Replacement series.

Additive Series: Mostly followed in India. One crop known as the base crops is sown at 100% rate. Other crops known as intercrops are also grown alongwith the base crop at a rate lower than cent percent.

Replacement Series: Mostly practiced in western countries. Both the crops are called component crops. Certain proportion of population of one component is sacrificed for growing the second crop.

1.4 INTERCROPPING CONCEPTS

Intercropping concepts include four important points:

- Spatial arrangement,
- Plant density,
- Maturity dates of the crops being grown, and
- Plant architecture.

1.4.1 *SPATIAL ARRANGEMENT*

There are at least four basic spatial arrangements used in intercropping.

Row Intercropping

Two or more crops are grown at the same time with at least one crop planted in rows.

Strip Intercropping

Two or more crops are grown together in strips wide enough to permit separate crop production using machines but close enough for the crops to interact.

Mixed Intercropping

Growing two or more crops together in no distinct row arrangement.

Relay Intercropping

Planting a second crop into a standing crop at a time when the standing crop is at its reproductive stage but before harvesting.

1.4.2 PLANT DENSITY

Plant density can be optimized by adjusting the seeding rate of each crop. It should be kept below the full rate to avoid overcrowding. The ratio of seed rate depends upon the ratio of yield needed from the intercrop.

1.4.3 MATURITY DATES

Its beneficial if the intercrops differ in their maturity dates or development periods. It reduces the competition between the two crops, It also helps in harvesting and separation of grain. Sorghum/pigeonpea intercrop is common in India. The sorghum matures in about four months. After the harvest of sorghum, pigeonpea flowers and ripens. Pigeonpea does not affect the sorghum yield (Willy, 1983).

1.4.4 PLANT ARCHITECTURE

Plant architecture is essential to balance the availability of certain factors such as sunlight availability. It can be observed in an intercrop of corn plants and beans and pumpkins.

1.5 ADVANTAGES OF INTERCROPPING

1. **Increasing production**

   The primary reason for the increase in intercropping practices around the globe include the increase in productivity of crops (Ghanbari and Lee, 2002). Odhiambo and Ariga, (2001) reported an increase in production with maize and beans intercrops in different ratios.

2. **Greater use of environmental resources**

   Due to the difference in the nutrient and resource requirement, intercropping leads to better utilization of environmental resources (Mahapatra, 2011).

3. **Reduction of pests, diseases and weeds damage**

   Soria *et al.*, (1975) reported that intercropping is effective in weed control with corn- cassava and beans- cassava intercrops. Fujita *et al.*, 1992 reported that intercropping does not always decrease pest or pathogen attack but most reports have shown that intercropping has a decreasing effect on pests and diseases.

4. **Stability and uniformity Yield**

   In case of solo cropping, natural calamities can lead to complete loss of the crop yield. But in case intercropping, failure of one crop still leaves the other crops unaffected in most of the cases. This ensures some return to the poor farmers.

5. **Improve soil fertility and increase in nitrogen**

   Leguminous plants show symbiotic relationship with *Rhizobium* bacteria which are able to fix atmospheric nitrogen into available nitrogen for plants uptake. This adds nitrogen, an essential nutrient to the soil which can be availed by other crops as well. Anil et al., 1998 and Fujita et al., 1992 reported that the nitrogen content in non-legume plants increased, due to the intercropping with leguminous plants.

6. **Reduced chemical use**

   Intercropping also reduces the cost of fertilizer application and pesticide requirements to the farmers.

7. **Overyielding**

   If the yield produced by the component crops grown in intercropping is larger than that of monoculture under similar conditions, then it is known as overyielding. Overyielding is calculated through the Land Equivalency

Ratio (LER) which is a measure of the land required to achieve intercrop yields with crops grown as pure stands.

8. ***Improvement of forage quality***

Cereal forages when grown with crops capable of increasing the protein content of the forage would have higher nutritional value.

9. ***Promotion of biodiversity***

Intercropping increases the biodiversity of the agroecosystem. Higher species richness may be associated with nutrient cycling characteristics that often can regulate soil fertility (Russell, 2002).

1.6 DISADVANTAGES OF INTERCROPPING

1. Yield may decrease due to the difference in their requirements.
2. Additional machinery cost required.
3. Improved technology might not be used efficiently.
4. Harvesting might be difficult.
5. Not adaptable to harsh conditions such as drought.
6. Very timely field operations are required.

1.7 INTERCROPPING : GLOBAL SCENARIO

Clover and grasses intercropping in pastures are extensively used in European farming but arable intercropping of main cereal crops is not so common these days. In the mechanized agricultural sector of Europe, North America, and some parts of Asia, intercropping is far less widespread (Horwith, 1985).

Example 1

- *Country :* Central America
- *Intercrops :* Corn, beans, and squash
- The corn has more height than the two crops.

- The beans climb up the corn stalks.
- The squash plants spread out along the ground.
- The shade discourages weeds from growing.

Example 2

- *Country :* Canada
- *Intercrops :* Corn and soyabean

Example 3

- *Country :* South Dakota
- *Intercrops :* Corn, soybeans, and spring wheat

Example 4

- *Country :* Latin America
- *Base crop :* Beans
- *Companion crop :* maize, potatoes, and other crops

Example 5

- *Country :* Africa
- *Base crop :* Cowpeas

Example 6

- *Country :* Colombia
- *Base crop :* Beans

1.8 INTERCROPPING IN INDIA

Due to diverse agro-climatic conditions in the country, a large number of crops are grown. Nearly 66 percent of the total cultivated area is under food grain crops (cereals and pulses). Commercial agriculture not only catered to the domestic market, but has also been one of the major sources of earning of foreign exchange for the country.

Inter Cropping in Cereals:

- *Base crop :* Sorghum (matures in 3 ½ to 4 ½ months
- *Companion crop :* pigeon pea (matures in 6-9 months depending on the genotype).
- *Base crop :* Maize
- *Companion crop :* Soybean or groundnut, black-gram or castor

Intercropping in Pulses:

- *Base crop :* pigeon pea,
- *Companion crop :* legumes such as black-gram and soybean, groundnut

Intercropping in cotton:

- *Base crop :* Cotton (slow growing crop)
- *Companion crop :* Any short duration and fast growing crops such as groundnut, black-gram, green-gram or cluster bean

Intercropping in Sugarcane:

- *Base crop :* Sugarcane, (period 80-90 days, planted in rows 0.8-1.0 m apart)
- *Companion crop :* Short duration crops maturing in 80-90 days as considerable space for inter-cropping is available. Eg:- Black-gram and soybean or the green manure, Dhaicha.

Intercropping in Coffee:

- *Area :* Karnataka and at the Araku Valley in Andhra Pradesh
- *Base crop :* Coffee
- *Companion crop :* Pepper

Cropping systems based on rice (*Oryza sativa* L.) are prevalent in the eastern part of India, which covers 43 per cent of rice area of the country. Rice-based systems are intimately connected with development of water resources. Chhattisgarh is popularly recognized as rice bowl of the country, as it is the principle crop of this state and about 69-70 percent of the net sown area is covered in *kharif* rice, while, 16.00 lakh hectares are cultivated under *rabi* season. With the increase in area under minor and major irrigation projects in Chhattisgarh, a number of crops can profitably be grown during the winter (*rabi*) and summer season followed by rice under irrigated conditions. Rice (*Oryza sativa* L.) is grown intensively during the rainy season, whereas, cereals like wheat (*Triticum aestivum* L.), mustard (*Brassica juncea*), sunflower (*Helianthus annuus* L.) and castor (*Ricinus communis*) as oilseeds, coriander (*Coriandrum sativum* L.), onion (*Allium cepa* L.), fenugreek (*Trigonella foenugraecum*) as spices and lentil (*Lens culinaris*) and chickpea, lathyrus as legume are the major *rabi* crops grown under irrigated conditions. Some of the districts of Chattisgarh have more than 35 percent net irrigated area. Therefore, to utilize the irrigation facilities in this area, there is need to diversify the existing cropping system and introduce some new high yielding profitable crops which can sustain and well suited under Chattisgarh agro-climatic conditions. This will not only enhance the socio-economic conditions of the farmers by providing employment for longer duration but also enable them to exploit the upcoming marketing and processing infrastructure in this area. Furthermore, development of improved technology with proper crop sequence plays a major role in getting maximum net return. Therefore, it becomes imperative to find the proper crop sequence. In order to generate useful information for such type of potential areas, an investigation was undertaken to study growth resource use and yield complementary of wheat-based intercropping system, *viz.* – wheat + lentil, mustard + lentil, sunflower + lentil, wheat + fenugreek, onion + coriander, castor + lentil under irrigated condition.

Keeping these point in view, a field experiment on the productivity and profitability of intercropping in *rabi* cereal, legume, oilseeds, and spices under rice (*Oryza sativa* L.) based cropping system was undertaken at Research cum Instructional Farm, Department of Agronomy I.G.K.V., Raipur C. G. during *kharif*, *rabi* season of 2009-2010 with the following objectives.

1. To evaluate the production potential and economic viability of different *rabi* intercropping under rice based cropping systems.
2. To identify suitable/ remunerative rice based cropping systems with vegetables and oilseeds intercrops.
3. To calculate the energy requirement, energy output and energy use efficiency of different rice based cropping systems.
4. To assess the effect of bio-intensive crops like oilseeds and pulses on the soil properties.

REVIEW OF LITERATURE

CHAPTER-II

REVIEW LITERATURE

In the era of shrinking resource base of land, water and energy, resource-use efficiency is an important aspect for considering the suitability of a cropping system. Since any modification to the existing system with a tendency to decline the productivity of rice crop will neither be sustainable nor acceptable to the farming community. Hence, Diversification and intensification of rice-based systems to increase productivity per unit resource is very pertinent. Choice of the component crops needs to be suitably maneuvered to harvest the synergism among them towards efficient utilization of resource base and to increase overall productivity. Research finding is available on the suitability of different cropping systems with high yielding varieties under rice-based cropping system are reviewed under the following heads.

2.1 Effect of cropping systems on

2.1.1 Growth and yield of rice

2.1.2 Total productivity of system including field crops

2.1.3 Total productivity of system including vegetables

2.1.4 Total productivity of system with intercrop

2.2 Effect of cropping systems on soil fertility status

2.3 Effect of cropping systems on weed dynamics

2.4 Water use efficiency

2.5 Economic viability

2.6 Employment generation, production and land utilization efficiency

2.7 Energetics

2.1 Effect of cropping systems on

2.1.1 Growth and yield of rice

Quayyam and Maniruzzaman (1996) reported that rice-groundnut - green gram system resulted in maximum number of effective tillers in rice (362m^{-2}), longest panicles (23.0 cm) and maximum number of grains panicle^{-1} (112), but the traits were similar to those of rice-groundnut-cowpea, rice-maize-green gram, rice-maize-cowpea and rice-sunflower- green gram systems. All these systems also produced higher grain and straw yields of rice, which may be attributed to the beneficial effect of legumes grown in summer season.

Singh and Sharma (2002) reported that the grain yield of rice showed significant variation under different cropping sequences and the maximum grain yield (16.51 t ha^{-1}) was found when sequenced in wheat-mung green manure-rice sequence. The rice yield was comparatively low in wheat-rice sequence. Singh and Tuteja (2000) found that the rice-potato recorded maximum rice equivalent yield followed by rice-mustard and rice-wheat cropping system.

Bastia *et al.* (2008) reported that rice-groundnut-green gram system resulted in maximum number of effective tillers in rice (362 m^{-2}), longest panicles (23.0 cm) and maximum number of grains panicle^{-1} (112). However, system productivity of rice-maize-cowpea was the maximum (15.98 t ha^{-1}) which was on a par with that of rice-maize- green gram (15.30 t ha^{-1}).

2.1.2 Total productivity of system including field crops

Parihar *et al.* (1995) reported that amongst different cropping sequences, rice – wheat cropping system gave the highest total grain production, followed by rice – rapeseed, under clay loam soils of Bilaspur, Chhattisgarh.

Settee and Gouda (1997) reported that system productivity in terms of rice-equivalent yield (REY) of rice-maize-cowpea was the maximum (15.98 t ha^{-1}), closely followed by that of rice-maize-green gram. Next in order was rice-field pea- sesame system. The winter crops mostly governed the REY of the systems, because rice was the base crop and contribution of summer crops was marginal. The contribution of winter crops to REY of rice-maize-cowpea, rice-maize-green gram and rice-field pea-sesame was 47, 48 and 44 percent, respectively. Other crops remaining the same, inclusion of summer cowpea as a vegetable crop increased the productivity of the systems and showed an edge over green gram and sesame.

Thakur et al. (1998) found that rice-sweet potato rotation recorded significantly better in terms of rice equivalent (52.6 q ha^{-1}) and productivity efficiency (17.94) followed by rice-wheat cropping system. Parihar *et al.* (1999) found that the rice-groundnut was more productive followed by rice-rice and rice-wheat. The lowest rice equivalent yield was obtained in rice-mustard sequence.

Choudhary *et al.* (2001) also reported greater productivity by replacing wheat in rice wheat system with vegetables like radish and potato. Kharub *et al.* (2003) reported that the maximum yield was obtained from rice – potato – sunflower (23.92 t ha^{-1}). The addition of short – duration potato crop between two main crops increased the productivity of the rice – wheat system. Kumar *et al.* (2005) observed that the rice-rice, rice-maize and rice-sunflower cropping sequences gave significantly higher rice equivalent yield

(11616, 11553 and 10868 kg ha^{-1}, respectively). Singh *et al.* (2005) reported that the rice equivalent yield (REY) was maximum in coarse rice-potato-sunflower (25.06 t ha^{-1}) followed by coarse rice-potato-late wheat (24.98 t ha^{-1}), basmati rice-potato-sunflower (17.71 t ha^{-1}), basmati rice-potato-late wheat(13.12 t ha^{-1}) as compared to traditional coarse rice-wheat (11.23 t ha^{-1}) system.

Yadav *et al.* (2005) reported that the maximum yield of rice in rice-wheat green manuring, where green manuring was taken in summer after wheat followed by coarse rice-mustard-sunflower sequence.

Singh and Singh (2005) reported that rice-onion gave the highest yield (118.97 q ha^{-1}) in the term of rice equivalent yield with maximum production efficiency (33.10 kg day^{-1} ha^{-1}).

Saroch *et al.* (2005) also reported more productivity by replacing wheat in rice – wheat system with vegetables. However, Gangwar *et al.* (2004) also noted higher stability of field crops in cereal – cereal or cereal – oilseed cropping systems. Bastia *et al.* (2008) reported that rice-sunflower-green gram registered the minimum rice equivalent yield (REY) of 11.66 t ha^{-1} among the three-crop sequences and rice-toria-fallow (8.36 t ha^{-1}) among the two-crop sequences.

2.1.3 Total productivity of system including vegetables

Diversification of existing rice based cereal/oilseed/pulse cropping system with vegetables has got great success in irrigated conditions. Inclusion of vegetables can not only increase the total productivity and net return from whole system but is also opens gateways to enhance employment generation and fulfill the demand of fresh vegetables.

Yadav *et al.* (2000) and Singh and Tuteja (2000) revealed that the rice-potato-cowpea provided the highest rice grain equivalent yield of 22.55 q ha^{-1} than rice-potato-okra (20.02 q ha^{-1}). Gangwar and Katyal (2001) found that the rice-potato-jute sequence yielded the highest during all the years, with mean yield of 16,936 q ha^{-1} $year^{-1}$ having system productivity of 67.47 kg day^{-1} ha^{-1} at Kalyani (West Bengal).

Similarly, Yadav *et al.* (2000) and Singh and Tuteja (2000) reported that sequences involving rice-tomato-poi was distinctly better than those over the years with mean yield equivalents of 26,680 kg ha^{-1} $year^{-1}$ at Bhubneshwar with the highest system productivity of (90.14 kg day^{-1} ha^{-1}). While, the lowest productivity was obtained in rice-mustard-rice sequence and in rice-mustard-ridge gourd sequence.

Choudhary *et al.* (2001) also reported that inclusion of oilseeds, vegetables, ornamental or fodder crops to diversify the existing rice-wheat system also helped in achieving higher rice-equivalent yield than with sequences having cereals and pulse crops.

Dhurandher *et al.* (2002) found that rice cabbage-onion-jowar recorded the highest rice equivalent yield (283.40 q ha^{-1}), production efficiency (82.04 kg day^{-1} ha^{-1}). Sharma *et al.* (2004) also reported that maximum yield (26.94 t ha^{-1}) was obtained from rice-potato-onion system followed by rice-potato-sunflower (23.92 t ha^{-1}). Bohra (2005) reported that rice-potato-green gram sequence was significantly and distinctly better than other system, *i.e.*, rice-lentil + mustard-cowpea (f), rice-wheat-green gram and rice-lentil- cowpea (f).

Manjunath and Korikanthimath (2004) observed that system productivity was highest with rice-brinjal system (11222 kg ha^{-1}) followed by rice-cowpea system (7681 kg ha^{-1}). Saroch *et al.* (2005) also reported more productivity by replacing wheat in rice-wheat system with vegetables. Due to high market price of mustard, rice-mustard-GM (81.96 q ha^{-1}) was comparable with existing rice-wheat cropping system (93.10 q ha^{-1}).

Urkurkar et *al.* (2008) found that the total productivity in terms of rice equivalent yield was significantly higher in rice-potato-cowpea cropping system (221.61 q ha^{-1}) than other systems. It was at par but produced much higher total productivity over rice-brinjal-green manure (181.16 q ha^{-1}) and rice-onion- green manure (160.63 q ha^{-1}). Rice based systems with field crops *viz.* rice-wheat- fallow, rice-mustard- green manure and rice-table pea-maize (fodder) produced almost similar rice equivalent yield with each other and these cropping systems were remained significantly lower than those systems which included vegetables and cash crops follow rice.

Singh *et al.* (2010) found that inclusion of potato in any of the above crop sequences proved beneficial in enhancing the productivity and profitability of the system. This may be attributed to the deep hoeing of the field because of ridge planting and hilling up, as well as the digging of potato tubers, which caused better soil aeration and weed free conditions for the Japanese mint, green gram and onion crops. Apart from this, potato was given heavy doses of fertilizers and perhaps did not utilize all the applied fertilizers, and later on, the following crops of onion, Japanese mint and green gram might have utilized the residual fraction.

2.1.4 Total productivity of system with intercrop

Improving resource utilization in time and space dimension is achieved through inter cropping. Singh and Singh (1983) reported that highest mean land equivalent ratio (1.27) was recorded in wheat and gram intercropping system, followed by wheat + pea (1.19) and wheat + lentil (1.10). Mandal *et al.* (1986) reported that intercropping of wheat and lentil generated a bonus yield that of wheat and chickpea gave an additional yield without any significant reduction in wheat yield.

Devi *et al.* (1997) studied that rice-chickpea-maize + cowpea fodder had highest productivity & net profit and the same may be recommended for sustainable cultivation under tarai region & alternative to rice – wheat sequence. Kumar *et al.* (2001) found that highest rice equivalent productivity (kg ha^{-1}) was obtained in rice-potato + onion, mustard + black gram system having productivity of 53.1 kg day^{-1} ha^{-1}. Similarly, Singh *et al.* (2001) found that rice- lentil-maize + fodder cowpea system gave significantly highest rice equivalent yield.

Singh *et al.* (2004) reported that the rice-lentil-maize + cowpea (f) sequence in flood-prone and rice-wheat-maize + cowpea (f) in semi-deep water situation gave significantly the highest rice equivalent yield (95.82 and 106.7 q ha^{-1}, respectively). Ganvir *et al.* (2004) reported that among the treatments, castor + groundnut (1:2) gave the highest castor and intercrop yield, total productivity, castor grain yield equivalent and gross monetary returns. Thakur *et al.* (2004) studied that Sunflower + chickpea (1:1) gave the maximum plant height (100 cm) and land equivalent ratio (1.27). Sunflower + linseed (1:1) gave the highest head size (12.5 cm) and grain yield

(1525 kg ha^{-1}). Sunflower + niger (1:1) had the highest number of seeds per head (279) and relative crowding coefficient (3.33). Sunflower + pea (1:1) while, sunflower + pea (1:2) and sunflower + linseed (1:2) gave the highest seed chaffiness (9.2%), sunflower equivalent yield (1101 kg ha^{-1}) and stem girth (5.0 cm), respectively.

Panwar and Rajbir (2004) found that the maximum *Brassica napus* yield equivalent (25.21 q ha^{-1}) was observed in *Brassica napus*+chickpea. The yield equivalents obtained under *Brassica napus* + toria and *Brassica napus*+ *Brassica juncea* combinations were significantly less (19.90 and 19.08 q ha^{-1}, respectively) than sole cropping of *Brassica napus* (21.27 q ha^{-1}).

Padmavathi and Raghavaiah (2004) found that the seed yield of castor (471 kg ha^{-1}) was adversely affected due to intercropping when compared to the sole crop of castor (748 kg ha^{-1}). The castor-equivalent yield was greater when castor was intercropped with cluster bean (1259; 2026 kg ha^{-1}) and cucumber (1536; 2050 kg ha^{-1}) either in 1 or 2 rows, respectively. Sharma *et al.* (2008) reported that among the 14 rice-based cropping systems tested, rice-potato–onion + maize relay cropping gave the highest mean rice-equivalent yield (30.66 t ha^{-1} $year^{-1}$), followed by rice-garlic - maize (30.35 t ha^{-1} $year^{-1}$) and rice-potato-onion (27.95 t ha^{-1} $year^{-1}$).

2.2 Effect of cropping systems on soil fertility status

Nambiar and Abrol (1989), Kumar and Yadav (1993) reported that continuously following the same system (rice – wheat sequence) has diverse effect on soil conditions, which ultimately reducing the productivity of the system. Singh and Prasad (1994) found that maximum nitrogen balance was recorded under rice-gram-green gram (128 kg N ha^{-1}) cropping sequence, whereas maximum potassium balance was under rice-maize, black gram (184 kg ha^{-1}). However, phosphorus-balance sheet showed a loss of in all the crop sequences with maximum loss under rice-potato-green gram (214 kg P ha^{-1}) sequence.

Bharadwaj and Omanwar (1994) reported that the highest net gain of K was observed with rice-fenugreek (+97.5 kg ha^{-1}) followed by rice-wheat (+76.2 kg ha^{-1}). The maximum gain of K by rice-fenugreek might be attributed to the direct addition of K to the available K pool of the soil besides the reduction of K fixation and release of K due to interaction of organic matter with clay.

Singh *et al.* (1996) and Setty and Gowda (1997) observed that the introduction of legume in the system increased soil organic carbon and available soil phosphorus. Singh *et al.* (1996) and Quayyam and Maniruzzaman (1996) found that the inclusion of legumes makes less demand on the soil resources and at the same time they have capacity to fix atmospheric nitrogen in their root nodules and helped in increasing the yield of succeeding rice crop. While, Thakur *et al.* (1998) reported that cropping system involving pulse crops in the winter (*rabi*) season had better soil fertility (N, P and K) status than those involving winter cereals. On the other hand,

Kumpawat (2001) studied that productivity of rice – wheat system had shown consistently declining trend. Inclusion of pulses, oilseeds & vegetables in the system has the more beneficial effect than cereal – after cereals. Sharma and Sharma (2002) observed that rice-berseem cropping system resulted in negative balance of nitrogen (144 kg ha^{-1}), phosphorus (23 kg ha^{-1})

and potassium (416 kg ha^{-1}). Rice-mustard-green gram cropping system also resulted in negative balance of 131 kg ha^{-1} N and 330 kg ha^{-1} K. Nitrogen and phosphorus balance was found positive in rice-wheat-green gram and rice-potato-green gram cropping system. Whereas, potassium balance was negative in these cropping systems.

Sharma and Sharma (2002) reported that the balance of P was positive in rice-wheat-green gram, rice-potato-sunflower, rice-garlic-maize, rice-marigold- maize + green gram, rice-fenugreek-maize and rice-sunflower-okra cropping system, and it varied from 6.30 kg ha^{-1} $year^{-1}$ in rice- fenugreek-maize to 28.20 kg ha^{-1} $year^{-1}$ in rice- marigold-maize + green gram cropping system. This shows that the P removed by the crops was less than that the applied to them. However, the other cropping systems showed a negative balance. The maximum deficit of P (31.40 kg^{-1} ha^{-1} $year^{-1}$) was observed in rice-berseem maize + cowpea (f) cropping system, indicating that the quantity of P applied to fodder crops was less than that removal from the soil.

Singh *et al.* (2004) found that available nutrients like nitrogen, phosphorus and potassium was improved due to legume included in cropping sequence. However, cereal and oilseed included in cropping sequences reduced the content of available nitrogen, phosphorus, potassium and organic carbon due to higher uptake and lower addition of nutrients in soil.

2.3 Effect of cropping systems on weed dynamics

The phenomenon of shift in density and dynamics of weed species is the results of manipulation of environmental factors, crop rotation and weed control practices. The continuous adaptation of the same crop rotation coupled with the use of particular weed management practices lead to shift in weed flora in time.

Gill and Brar (1985) in Ludhiana found that rice-wheat, maize-wheat and maize-potato-wheat rotation indicated that continuous use of herbicide in the rice-wheat rotation suppressed the wild oat population in wheat and encourage *Phalaris minor*, similarly maize-wheat rotation showed gradual build up of wild oat in wheat. Tiwari *et al.* (1990) noticed that continuous cropping of rice-wheat-green gram resulted into more number of aerial shoots (270 m^{-2}) and tuber count (953 m^{-2}) of *Cyperus rotundus* as compared to rice-wheat cropping system.

Srinivasan *et al.* (1992) noticed the shifting of weed flora from annual (*Echinochloa spp.*) to perennial (*Paspalum distichum* and *Massilea minuta*) in rice-greengram sequence. Bhan and Kumar (1998) found that the changing sequence from rice-wheat and rice-potato to any other sequence not involving rice in kharif tends to reduce *Phalaris minor* population in wheat.

Gangwar and Ram (2005) revealed that population of *Phalaris minor* and dry weight was significantly less (53.6 %) in wheat under rice-vegetable pea-wheat-green gram sequence. It was closely followed in the system involving rice-mustard-green gram, rice-wheat-green gram and rice-wheat-green gram sequence managing reduction to the tune of 46.4%. Similarly in rice-berseem (fodder), rice-wheat, rice-wheat sequence, the weed control was up to 38.5% over continuous rice-wheat sequence. While, it was higher in rice-wheat sequence (24.6 gm^{-2}). Singh and Singh (2005) observed that weed density rice under rice-wheat, rice-pea-rice cropping system, *Echinochloa crusgalli* was dominated after rice-wheat and rice-pea-rice sequence, where as *Cyprus difformis* was dominant after rice-pea-rice sequence.

2.4 Water use efficiency

Mandal *et al.* (1986) studied that growing of cereals with pulses & oilseeds endowed with different root systems and helped towards better extraction of soil moisture from different layers of the soil, increase water use & water use efficiency as well as intercepting more solar energy. They also reported that intercropping can give higher yield than when grown as sole crops.

Singh and Prasad (1994) recorded highest frequency of irrigation under rice-wheat-green gram with and minimum in rice-mustard-green gram. Parihar *et al.* (1999) reported that total water use was maximum in rice-rice crop sequence followed by rice-groundnut. The highest water expense under rice-rice system was mainly due to substantially high water requirement of summer rice. The lowest water use among the different cropping sequences was in rice –chickpea system (10.25 cm). The water use efficiency (WUE) was the highest in rice –chickpea system followed by rice-sunflower. The lowest WUE was obtained in rice-rice cropping sequence (45.78 grains ha^{-1} cm). Similarly, Kumar *et al.* (2005) reported that field water supply was higher in rice-rice cropping sequence (2,800 mm) and water use efficiency was lowest (4.15 kg mm^{-1}). On the other hand, rice-maize and rice-sunflower cropping sequences recorded higher water use efficiency (7.22 and 7.30 kg mm^{-1}), respectively.

Singh *et al.* (2001) reported that the WUE was significantly highest in rice-lentil-maize+cowpea (83.83 kg grain ha^{-1} cm) followed by rice-wheat-maize+cowpea (f). The lowest water use efficiency was recorded in rice-sunflower-maize+cowpea (f) cropping sequence. Singh *et al.* (2004) reported that WUE was invariably highest in rice-lentil-maize + cowpea (f) in both the situation i.e.83.8 kg ha^{-1} cm^{-1} in flood prone and 71.77 kg ha^{-1} cm^{-1} in semi-deep water situation, respectively.

Gangwar and Ram (2005) opined that if the total irrigation water requirement of different system is concerned, higher irrigation water of 117.4 ha^{-1} cm was required in rice-berseem fodder (one year)-rice-wheat-rice-wheat sequence while, it was lowest and similar in sequence *viz.* pigeon pea-wheat-rice-wheat-rice-wheat, and rice-mustard-green gram-rice. Wheat-green gram-rice-wheat-green gram required about 100 ha^{-1} cm of irrigation water.While, Kumar *et al.* (2008) observed that rice – potato – green gram sequence was the most efficient with respect to water-use efficiency, followed by rice - onion. Berseem may be taken as a break crop successfully for reducing the weed problem effectively. Sharma *et al.* (2008) reported that the highest water-use efficiency (37.01 kg rice-equivalent yield ha^{-1} mm^{-1}) was recorded with rice–garlic–maize system.

2.5 Economic viability

Connor *et al.* (2003) observed that rice - vegetable pea – wheat - green gram sequence produced 27.91% higher wheat equivalent yield than rice - wheat system. Diversification or intensification of rice - wheat system, once in 3 years, improved the net returns when all the crops (except rice) were grown on raised-bed in a system approach.

Jat and Singh (2003) observed that furrow-irrigated raised-bed planting of wheat with 2 rows bed^{-1} and simultaneous sowing of berseem in furrows with multiple cuttings gave the highest wheat-equivalent yield (7.03 t ha^{-1}) and per day productivity (57.07 kg ha^{-1} day^{-1}). This system also recorded the highest net returns (Rs. 33265 ha^{-1}).

Nayak *et al.* (2003) found that the highest rice equivalent yield (14116 kg ha^{-1}) and gross returns (Rs 74815 ha^{-1}) were recorded from rice-tomato-okra, followed by rice-Indian mustard-groundnut (11776 kg ha^{-1} and Rs. 62423 ha^{-1}). However, the highest net returns were recorded with rice-indian mustard-groundnut (Rs 35539 ha^{-1}), followed by rice-tomato-okra (Rs 32031 ha^{-1}) with benefit: cost ratio of 2.32 and 2.05 respectively. Rice-chili, a two-crop system, also gave comparable net returns of Rs 27524 ha^{-1} with high benefit: cost ratio of 2.13, although land uses efficiency was lowest.

Prasad *et al.* (2003) reported that the highest chickpea equivalent yield (31.19 q ha^{-1}), land equivalent ratio (1.15), gross income (Rs. 43572.4 ha^{-1}) and net profit (Rs. 29 216.4 ha^{-1}) were recorded with intercropping of chickpea and mustard cv. Urvashi.

Bhagat and Dhar (2003) studied that the enhancement in the productivity and economics of the traditional rice-wheat system by modifying it to maize-wheat, groundnut-wheat and soyabean-wheat system. Among the 3 modified systems. The highest net return (Rs. 20533 ha^{-1}) was recorded from groundnut-wheat system with 112% increase in yield over the traditional rice-wheat system.

Sharma *et al.* (2004) found that among the different rice-based crop sequences, the rice-potato-onion fetched the highest net returns (Rs 65573 ha^{-1}) and production efficiency (81.6 kg ha^{-1} day^{-1}), followed by the rice-potato-sunflower having corresponding values of Rs. 61533 ha^{-1} and 80.6 kg ha^{-1} day^{-1}. The highest benefit:cost ratio of 1:47 was noted in rice-berseem-maize + cowpea both grown for fodder. Rice-maize + green gram system had the highest land-use efficiency (96.8%). Gangwar and Ram (2005) also reported higher benefit: cost ratio (2.4) by this crop sequence compared with other crop sequences.

Sarkar *et al.* (2004) found that an intercropping of 100% lentil+25% linseed in 5:1 ratio gave as high intercrop lentil yield (1050 kg ha^{-1}) as sole lentil (1300 kg ha^{-1}) with additional yield of linseed (820 kg ha^{-1}) and had the highest lentil-equivalent yield (1767 kg ha^{-1}), net return (Rs 17632 ha^{-1}), land-equivalent ratio (1.66), and benefit: cost ratio (2.65), and proved the most productive intercrop stand among all the systems.

Padmavathi and Raghavaiah (2004) found that the net returns were greater when castor was intercropped with cluster bean (Rs.6548 and Rs.12611 ha^{-1}) and cucumber (Rs.8791 and Rs.11186 ha^{-1}) either in 1 or 2 rows, respectively. Seed or vegetable yields of the intercrops were greater in their pure stands, followed by when these crops were sown in 2 rows and 1 row as intercrops with castor.

Kumar *et al.* (2005) found that among the 8 cropping sequences, rice-rice, rice-maize and rice-sunflower cropping sequences gave significantly higher rice-equivalent yield (11616, 11553 and 10868 kg ha^{-1}, respectively) and gross returns (Rs. 51745, 51464 and 48291 ha^{-1}, respectively). However, net returns and benefit: cost ratio were higher in rice-maize (Rs. 21884 ha^{-1} and 1.74, respectively) and rice-sunflower (Rs. 21321 ha^{-1} and 1.79, respectively) cropping sequences. Similarly, these cropping sequences also recorded higher energy use efficiency (6.4 and 5.1, respectively), energy productivity (227 and 238 g MJ^{-1}, respectively) and water use efficiency (7.22 and 7.50 kg mm $^{-1}$, respectively) and were more sustainable than rice-rice and other rice-based cropping sequences.

Kumar *et al.* (2008) observed that rice - wheat (normal sown) and rice - berseem, gave significantly higher net returns than rice - oat and rice - wheat (zero till) sequences. They also observed that the highest net returns of Rs. 43,100 ha^{-1} $year^{-1}$ was provided by rice – potato – green gram sequence, followed by rice - onion (Rs. 36,400 ha^{-1} $year^{-1}$), and both the sequences gave significantly higher net return than other crop sequences.

Bastia *et al.* (2008) also reported that returns per Re invested were the highest for rice-field pea-sesame system (Rs 1.94), which were on a par with that of rice-maize-cowpea and rice-maize-green gram systems (Rs 1.85 and 1.83, respectively). Sharma *et al.* (2008) reported that the highest net return of Rs 96,581/ha/year were realized from rice-garlic-maize, which were on a par with that of rice-potato-onion + maize relay cropping (Rs 92,837 ha^{-1} $year^{-1}$). However, the benefit: cost ratio was highest (1.73) in rice-berseem– maize + cowpea, both grown for fodder.

Urkurkar *et al.* (2008) reported that the maximum productivity was obtain under rice-potato-cowpea system but amongst all the systems, rice-brinjal-GM was identified to be distinctly superior and more economically viable in terms of net return (Rs. 83482 $^{-1}$ ha) and benefit:cost ratio (3.09) than other systems based on four year mean data. The higher cost involved in potato seed and other intercultural operations increased the cost of cultivation of rice-potato-cowpea system. Thus, rice-potato-cowpea system was in second rank in order of economic merit closely followed by rice-onion-GM. These systems also showed higher value of relative economic efficiency and have potential to replace existing systems i.e. rice-wheat or rice-mustard.

While, Singh *et al.* (2010) found that the rice-potato-Japanese mint and rice–potato–onion crop sequences are more productive and economically viable as they also fetched more net returns per unit area for time invested, and can be a better option for the farmers of the Central Plains Zone of India. These two cropping sequences were also judged as most sustainable by the farmers in their local agricultural production system.

2.6 Employment generation, production and lands utilization efficiency

Chauhan *et al.* (2001) obtained the highest production efficiency in rice – wheat crop sequence (46.95 kg ha^{-1} day^{-1}), followed by pigeon pea - wheat , rice – wheat (46.31 kg ha^{-1} day) and lowest in maize - vegetable pea – wheat , rice – wheat – green gram and rice – wheat – green gram (41.6 kg ha^{-1} day^{-1}).

Dhurandher *et al.* (2002) found that among the various cropping sequences, rice-cabbage-onion-jowar recorded the highest production efficiency (82.04 kg ha^{-1} day^{-1}) and land use efficiency (94.66%).

Sharma *et al.* (2004) also reported that intensification through inclusion of vegetable and leguminous crops increased the production and land use efficiency. The increase in employment generation in rice - potato – green gram and rice - onion system improved the profitability, but in rice - wheat (transplanted) system it did not give monetary advantage over the existing rice - wheat system. The highest economic efficiency (Rs.152 $ha^{-1}day^{-1}$) and water-use efficiency (20.3 kg^{-1} ha^{-1} mm^{-1}) were obtained in rice - onion system. Rice - berseem sequence was found most efficient in terms of

N-use efficiency (80.21 kg grain $^{-1}$ kg N) and third in production efficiency and land use efficiency. Late sowing of wheat after long-duration rice showed lower values of nitrogen and water-use efficiencies, which were very close to the respective minimum values obtained in rice-oat system. They found that the production efficiency was maximum in rice-potato-cowpea (70.4 kg^{-1} ha^{-1}day^{-1}) cropping sequence whereas, the lowest production efficiency (27.5 kg^{-1} ha^{-1}day^{-1}) was noted in rice-mustard-green manure as well as in rice-table pea-maize (fodder) sequence mainly due to less total productivity of former cropping sequences with rather longer duration.

Dungani *et al.* (2005) reported that the highest (9689 kg ha^{-1}) total production in terms of paddy equivalent yield and production efficiency (26.35 kg day $^{-1}$ ha$^{-1)}$ was obtained under rice-wheat-green gram sequence. The next best sequence was rice-sorghum (f)-groundnut with equivalent of 9609 kg ha^{-1}and production efficiency 26.3 kg day $^{-1}$ ha^{-1}. While the lowest total production (6537 kg ha^{-1}) and productivity (17.91 kg day $^{-1}$ ha^{-1}) was recorded in paddy-pigeonpea sequence.

Sharma *et al.* (2006) found that the rice-potato-onion + maize relay cropping have the highest rice equivalent yield (32.19 t ha^{-1} year $^{-1}$) and production efficiency (88.2 kg day $^{-1}$ ha^{-1}) followed by rice-potato-onion having corresponding values of 29.08 t ha^{-1} year $^{-1}$and 79.7 kg day $^{-1}$ ha^{-1}, respectively. Rice-garlic- maize was third in order, which produced rice equivalent yield of 27.35 t ha^{-1} year $^{-1}$ and production efficiency of 74.9 kg day $^{-1}$ ha^{-1}.

Kumar *et al.* (2008) reported that rice-potato-greengram sequence was found the most efficient for production (18.1 t^{-1} ha^{-1} year^{-1}), employment generation (1.18 man days^{-1} ha^{-1}day^{-1}), monetary return (Rs 3,180/ha/year) and water-use efficiency (20.1 kg^{-1} ha^{-1} mm^{-1}), followed by rice-onion. Berseem may be taken as a break crop successfully for reducing weed problem (weed-control efficiency 88.7%) in continuous rice-wheat system without any monetary loss. Rice - berseem sequence was also found the most efficient in terms of nitrogen-use efficiency (80.2 kg grain^{-1}kg N).

Kumar *et al.* (2008) also observed that rice – potato – green gram and rice-onion sequences gave 57.4 and 55.9 kg^{-1} ha^{-1} day^{-1} higher production efficiency compared with 45.1 kg^{-1} ha^{-1} day^{-1} in rice - wheat sequence. The land-use efficiency (86.3%) and employment-generation efficiency (Rs.1.18 days ha^{-1} day.) was found highest in rice – potato - greengram sequence due to intensification of this system.

Bastia *et al.* (2008) reported that Cropping systems having maize as a component crop expressed higher production efficiency (Rs. 124 to 128 ha^{-1} day^{-1}), water-use efficiency (Rs 203 to 213 ha^{-1}cm) and energy intensiveness (11.23 to 12.56 MJ Re^{-1}) in economic terms. Rice-maize-cowpea was the most productive, sustainable, resource-use efficient and remunerative cropping system. Further they also reported that production efficiency (50.73 kg REY ha^{-1} day^{-1}), employment generation (550 man days/ha) and net returns (Rs 40,415 ha^{-1}) were maximum in rice-maize-cowpea followed by rice-maize-green gram system.

Singh *et al.* (2010) studied that production efficiency values in terms of kg ha^{-1} day^{-1} was highest under the rice–potato–onion cropping sequence crop (87.25), closely followed by rice-potato-japanese mint (79.76), as compared to the lowest value of 41.16 under the rice–wheat–japanese mint cropping sequence. Land utilization efficiency values under different crop sequences was highest (94.5%) in the rice–wheat–japanese mint cropping sequence as compared to 90.7% and 91.5% under rice–potato–onion and rice–potato–japanese mint crop sequences, respectively.

2.7 Energetics

Padhi (1993) found that the maximum energy was utilized under rice-potato-okra system due to two high-energy use crops like potato (28.22 MJ x 10^3 ha^{-1}) and okra (15.66 MJ x 10^3 ha^{-1}), whereas rice-garden pea-cowpea utilized minimum energy due to inclusion of two low energy use crop like garden pea (8.26 MJ x 10^3 ha^{-1}) and cowpea (7.55 MJ x 10^3 $ha^{-1)}$. Rice-garden pea-okra recorded the lowest energy output and energy input : output ratio due to minimum energy production by the component crops like rice (44.98 MJ x 10^3 ha^{-1}) and garden pea (9.31 MJ x 10^3 ha^{-1}). Whereas, among different intercropping systems, Malik *et al.* (1993) reported that wheat + lentil and wheat + mustard gave 20 and 30%, respectively more energy output than wheat alone. Significant improvement in energy productivity under wheat + lentil was because of more grain production and that under wheat + mustard intercrop system was owing to much higher energy density within a unit mass of mustard compared to wheat and grain legumes.

Parihar *et al.* (1999) reported that rice-rice system required the highest energy input (27.35 x 10^3 MJ ha^{-1}) while, it has lowest in rice-chickpea (17.7 x 10^3 MJ ha^{-1}). However, rice-rice system produced the highest output energy followed by rice-groundnut.

Gangwar and Katyal (2001) in Kalyani (W.B.), observed the highest energy production (31.45 x 10^6 K cal ha^{-1}) on actual crop basis in rice-wheat-green manure, while at Bhubaneswar rice-potato-sesame (36.16×10^6 K cal ha^{-1}) sequence recorded the highest energy production. While, Singh *et al.* (2001) reported that rice-pea-maize + cow pea (f) system gave the highest energy output (907.75 MJ $\times 10^3$ ha^{-1}) and energy use efficiency (29.52) were obtained in rice-mustard-maize + cow pea (f) sequence.

Singh *et al.* (2004) reported that rice-mustard-maize + cowpea (f) recorded the highest energy output (907.08 and 950.0×10^3 MJ ha^{-1}) and energy use efficiency (29.52 and 33.37). The lowest energy output and energy use efficiency were recorded with rice- pea-maize + cow pea (F) cropping system. Sharma *et al.* (2004) reported that rice-potato-onion required the highest energy inputs (6990×10^3 MJ ha^{-1}) resulting in the lowest energy use efficiency (2.46), whereas rice-potato-sunflower production recorded the highest output energy (205.71×10^3 MJ ha^{-1}).

Kumar *et al.* (2005) reported that rice-rice cropping sequence required higher energy input (75.036 MJ ha^{-1}) but the energy use efficiency and energy productivity were lowest (4.2 and 155 g MJ^{-1}). On the other hand, Yadav *et al.* (2005) reported that maximum calorific value was found in maize potato-sunflower crop sequence followed by maize-potato-wheat and rice-wheat-green manuring.

Sharma *et al.* (2008) reported that the maximum energy output (61,155 K calories ha^{-1} $year^{-1}$) was found in rice-potato-onion + maize relay cropping system, followed by rice-maize + potato intercropping (57,996 K-calories ha^{-1} $year^{-1}$), rice-potato-sunflower and rice-potato-onion systems. This indicates that these cropping systems have high value of high-quality produce. The cropping systems in which potato and onion crops were not included i.e. rice-wheat-maize gave lower energy output with lowest energy output under rice-cabbage- okra cropping system.

MATERIALS AND METHODS

CHAPTER-III

MATERIALS AND METHODS

This chapter deals with the concise description of the materials used and the techniques adopted during the course of investigation. The present investigation entitled **"Productivity and profitability of intercropping in *rabi* cereal, legume, oilseeds and spices under rice (*Oryza sativa* L.) based cropping system"** was conducted at the Research cum Instructional Farm, I.G.K.V., Raipur (Chhattisgarh) during the *kharif* and *rabi* season, 2009-10.

3.1 Geographical Situation

The experiment was conducted at the Research cum Instructional Farm, I.G.K.V., Raipur (Chhattisgarh). Geographically, Raipur is situated in the centre of Chhattisgarh and lies between 21°4'N latitude and 81°39'E longitude with an altitude of 314 metres above the mean sea level.

3.2 Climatic Condition

The general climate of the experimental site is classified as sub-humid with hot summer and mild winters. It comes under the Chhattisgarh plains Agro-climatic sub zone of seventh Agro Climatic Region of India i.e. Eastern Plateau and Hills. It has an annual average rainfall of 1320 mm (based on 80 years mean), nearly 85 per cent of the annual rainfall is received from third week of June to mid of September. The maximum temperature raises upto 45^0C during summer season, whereas the minimum temperature falls to 5-6 ^0C during winters. Relative humidity is normally higher from June to September and thereafter declines in winters.

3.3 Weather condition during crop growth

The rainfall was favorable and well distributed during *kharif* season 2009. The total rainfall was 1054.6 mm. Out of this, rice crop received 962.6 mm during its growth period

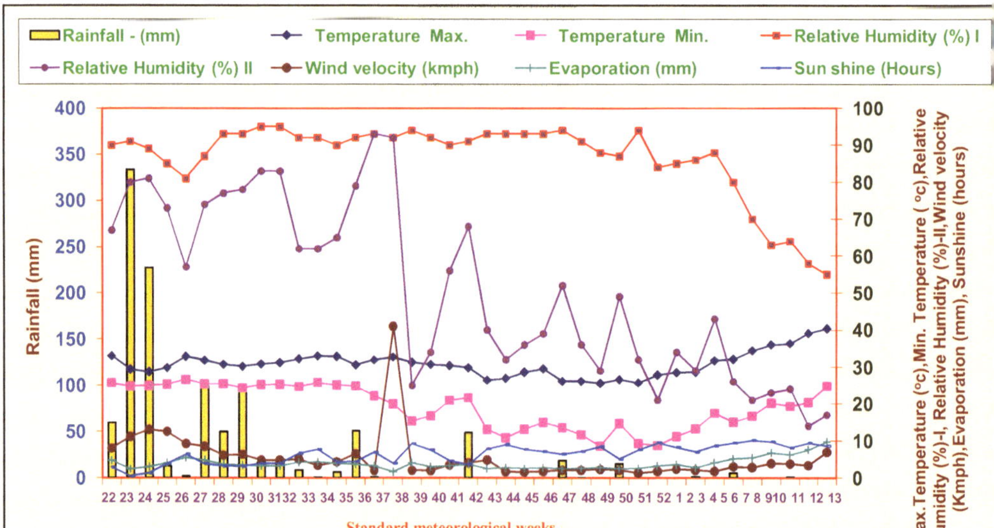

Fig 3.1: Weekly Meteorological parameters during crop growth period
(01,July 2009-30, March 2010)

and 92.0 mm rainfall was received by the *rabi* crops. The weekly maximum temperature ranged between 28.7^0C to 43.3^0C in *kharif* and between 25.6^0C to 39.7^0C during *rabi*. The minimum temperatures faced by the rice crop were 15.3^0C to 30.9^0C. While, the *rabi* crops enjoyed the minimum temperatures from 8.6^0C to 21^0C.

The monthly average maximum relative humidity for different months varied from 52.3 to 93.0 % whereas, the monthly minimum relative humidity varied between 19.7 to 80.0 %. Weekly averages of meteorological observation recorded during the course of investigation are presented in Appendix-I and depicted through Fig.3.1.

3.4 Cropping history of the field

The experimental field was under cultivation since many years and from 2003-04 to 2008-09, rice based seven cropping systems were taken in the same field since last 6 years.

3.5 Physico-chemical properties of experimental Soil

In order to evaluate the nutrient status of soil, ten samples were collected randomly up to 20 cm depth from five places to determine the physico-chemical properties of the soil. The procedure adopted for analysis and values obtained are given in Table 3.1. The soil of the experimental site is characterized by Silty clay (*Inceptisols)* in texture and locally known as "*Matasi*". The soil was neutral in reaction and medium in fertility having low nitrogen, medium phosphorus and potash.

Table 3.1: Physico-chemical properties of the experimental site

No.	Particulars	Values	Rating	Methods used
A. Physical properties				
1.	Mechanical composition			
	Sand (%)	23	Silty clay (*Inceptisols*)	International pipette method (Black, 1965)
	Silt (%)	46		
	Clay (%)	31		
B. Chemical composition				
1.	Organic carbon (%)	0.49	Medium	Walkey and Black's rapid titration method (Black, 1965)
2.	Available N (kg ha^{-1})	248	Medium	Alkaline permanganate method (Subbiah and Asija, 1956)
3.	Available P (kg ha^{-1})	18.6	Medium	Olsen's method (Olsen, 1954)
4.	Exchangeable K (kg ha^{-1})	262	Medium	Flame photometric method (Jackson, 1967)
5.	pH (1:2.5, Soil : water)	7.36	Neutral	Glass electrode pH meter (Piper, 1967)
6.	Electrical conductivity (m mhos m^{-1} at 25°C)	0.20	Normal	Solubridge method, (Black, 1965)
C. Physical composition				
3.	Bulk density (Mg m^{-3})			
	0-15 cm, soil depth	1.12		Soil Core Method (Black, 1942)
	15-30 cm, soil depth	1.28		

3.6 Experimental details

During *kharif*, rice was taken as base crop with three varieties and eight different cropping systems were followed during *rabi*. The experiment was laid out in randomized block design with 3 replications. Details of the treatments are presented in Table 3.2 and layout plan is depicted through Fig.

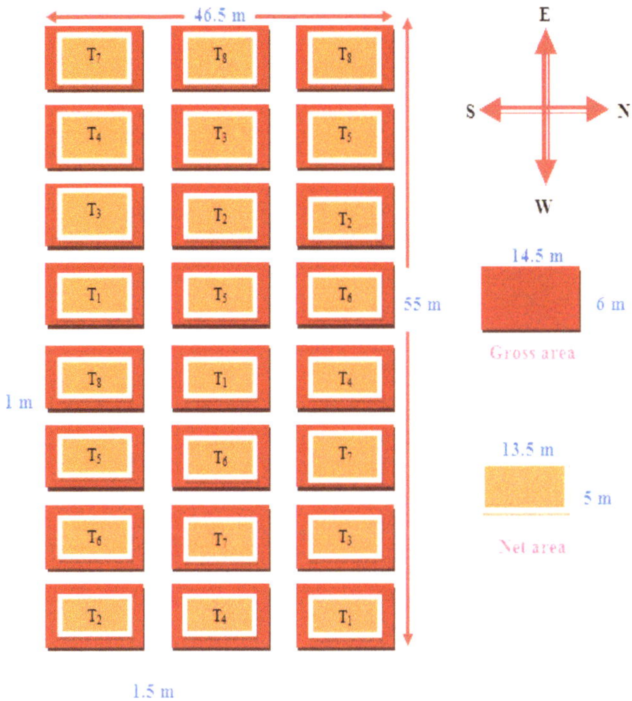

Notation	Kharif	Rabi
T1	Rice (MTU-1010)	Wheat
T2	Rice (MTU-1010)	Castor + Lentil (1:3)
T3	Rice (MTU-1010)	Mustard + Lentil (1:2)
T4	Rice (Mahamaya)	Sunflower + Lentil (1:3)
T5	Rice (Mahamaya)	Wheat+ Fenugreek(1:1) skip row
T6	Hybrid rice (Indira Sona)	Wheat + Lentil (1:1) skip row
T7	Hybrid rice (Indira Sona)	Onion + Coriander (3:1)
T8	Rice (MTU-1010) (Dummy treatment)	Wheat

Design	:	Randomized Block Design (RBD)
Treatment	:	Eight
No. of replication	:	Three

Fig. 3.2 : Plan of layout

Table 3.2: Treatment details of experiment

Tr	Kharif	Rabi
T_1.	Rice 'MTU 1010'	Wheat
T_2.	Rice 'MTU 1010'	Castor + Lentil (1:3)
T_3.	Rice 'MTU 1010'	Mustard +Lentil (1:2)
T_4.	Rice 'Mahamaya'	Sunflower + Lentil (1:3)
T_5.	Rice 'Mahamaya'	Wheat +Fenugreek (1:1 skip row)
T_6.	Hybrid Rice 'Indira Sona'	Wheat + Lentil (1:1 skip row)
T_7.	Hybrid Rice 'Indira Sona'	Onion + Coriander (3:1)
T_8.	Rice 'MTU 1010'	Wheat (Dummy treatment)

The present study was a fraction of an on-going long term experiment of All India Coordinated Research Project on Integrated Farming Systems (AICRP on IFS) at Raipur, which has been conducted with eight rice based triple cropping systems during all three seasons i.e. *kharif*, *rabi* and summer since 2009-10 (Table 3.3). The present study was based on *kharif* and *rabi* seasons' crops. The details of original experiment of AICRP on IFS are given below:-

Table 3.3: Details of original experiment

S.N.	Kharif	Rabi	Summer
T_1	Rice 'MTU-1010'	Wheat	Cowpea
T_2	Rice 'MTU-1010'	Castor+lentil (1:3)	Cowpea (fodder)
T_3	Rice 'MTU-1010'	Mustard+lentil (1:2)	Cowpea
T_4	Rice 'Mahamaya'	Sunflower+lentil (1:3)	Cowpea (veg.)+ residue incorporation
T_5	Rice 'Mahamaya'	Wheat+fenugreek (1:1 skip row) wheat in furrows and fenugreek(seed purpose) on beds	Cowpea (fodder)
T_6	Hybrid rice 'Indira Sona'	Wheat+lentil (1:1 skip row) wheat in furrows and lentil on ridges	Cowpea(veg.) in furrow+residue incorporation
T_7	Hybrid rice 'Indira Sona'	Onion+Coriander (green leaf and seed) (3:1)	Cowpea
T_8	Rice 'MTU-1010'	wheat	Cowpea

3.7 Test crops

3.7.1 Rice (*Oryza sativa* L.)

Three rice varieties, one each of early duration (MTU 1010) and medium duration (Mahmaya) and one hybrid (Indira sona) were sown as test crop in *kharif*.

Rice cv. 'MTU 1010' is an early duration variety and takes 112 days for maturity. It is a dwarf variety which is resistant to brown aphids.

Rice cv. 'Mahamaya' is a medium duration (128-130 days) high yielding semi dwarf variety. It is released from I.G.K.V., Raipur and resistant to gall midge.

Rice cv. 'Indira sona' is a hybrid and released from IGKV, Raipur was sown in *kharif* season. It is semi dwarf, matures in 125-130 days and having medium slender grain with good yield potential.

3.7.2 Wheat (*Triticum aestivum* L.)

Wheat cv. 'GW-273', has an average plant height of 95-100 cm with medium to profuse tillering. It takes 100-120 days to mature.

3.7.3 Castor (*Ricinus communis*)

Castor cv. 'DCS-9' (Joyti) is an early duration hybrid variety, which matures in 120-130 days.

3.7.4 Lentil (*Lens culinaris* MediK.)

Lentil cv. 'K-75' is bold seeded variety. It takes 120-125 to mature. It is resistant to wilt and rust.

3.7.5 Mustard (*Brassica juncea*)

Mustard cv. 'Pusa-Bold' is a derivative of Varuna x BLC-1780 released in 1984 by Central sub-committee of variety release. It is of 110-120 days duration with a plant height of 170-180 cm.

Plate 1: View of sunflower + lentil

Plate 2: View of wheat + fenugreek

3.7.6 Sunflower (*Helianthus annuus* L.)

Sunflower cv. 'Jawalamukhi', is a hybrid and having high yield potential. It is suitable for input intensive agriculture with average 44% oil content.

3.7.7 Fenugreek (*Trigonella foenumgraceum*)

Fenugreek cv. 'PEB' is a popular and widely cultivated variety. Its shoots are erect standing and plants are quick growing.

3.7.8 Onion (*Allium cepa* L.)

Onion cv. 'Nasik red' is popular and widely cultivated variety.Bulbs are medium to large-sized. It takes 120-130 days to mature.

3.7.9 Coriander (*Coriandrum sativum* L.)

Coriander cv. 'Hariyali', having erect growth habit with dense dark green foliage and medium-sized grains. The green leaf of coriander is harvested for fragrance in meals.

3.8 Experimental details and cultural operations

Details of recommended dose of rice and different *rabi* crops (including intercrops), sowing and harvesting dates are given in Table 3.4.

Table 3.4 Experimental details of fertilizer doses and sowing and harvesting dates

S. No.	*Kharif*			*Rubi*		
	Rice variety	**$N:P_2O_5:K_2O$ (kg ha^{-1})**	**Transplanting/ Harvesting**	**$N:P_2O_5:K_2O$ (kg ha^{-1})**	**Seeding**	**Harvesting**
1.	Rice (MTU 1010)	80:60:40	23.07.09 01.11.09	Wheat -100:60:40	20.11.09	Wheat-09/03/10
2.	Rice (MTU 1010)	80:60:40	23.07.09 01.11.09	Wheat-100:60:40 Lentil-30:50:30	20.11.09	Wheat-09/03/10 Lentil-05/03/10
3.	Rice (MTU 1010)	80:60:40	23.07.09 01.11.09	Castor-80:50:30 Lentil-30:50:30	20.11.09	Castor-28/03/10 Lentil-05/03/10
4.	Rice (MTU 1010)	80:60:40	23.07.09 01.11.09	Sunflower-80:50:30 Lentil-30:50:30	20.11.09	Sunflower-26/03 Lentil-05/03/10
5.	Rice(Mahamaya)	100:60:40	23.07.09 01.11.09	Mustard -80:50:30 Lentil-30:50:30	20.11.09	Mustard-15/03/1 Lentil-05/03/10
6.	Rice(Mahamaya)	100:60:40	23.07.09 01.11.09	Onion-80:60:100 Coriander-60:40:30	20.11.09 23.11.09	Onion-25/03/10 Coriander- Vegetable-20/1/1 Seed-23/03/10
7.	Rice(Indira sona)	120:60:40	23.07.09 01.11.09	Fenugreek-40:40:30 Wheat- 100:60:40	20.11.09	Fenugreek-16/03 Wheat-09/03/10
8.	Rice(Indira sona)	120:60:40	23.07.09 01.11.09	Wheat -100:60:40	20.11.09	Wheat-09/03/10

Date of rice nursery -26.06.09, transplanting of onion-23.11.09

3.9 Seed treatment

Seeds of rice and different *rabi* crops treated with carbendazim @ 2.5 g kg^{-1} seed and then sown in each bed and covered with thin layer of soil for rice nursery raising as well as for direct seeding of rabi crops.

3.10 Transplanting of rice

For rice transplanting, the nursery was uprooted and transplanted into a well puddle experimental field by maintaining a spacing of 20 x 10 cm. Two to three seedlings were transplanting at each hill on a shallow depth.

3.11 Cultural Schedule

The field was prepared by cross ploughing with tractor drawn mould board plough followed by harrowing and planking. Subsequently, layout of *kharif,* and *rabi* experiment was carried out in the field during respective seasons. Details of different cultural operations in rice and *rabi* crops are given in Table 3.5.

3.12 Harvesting and threshing

Harvesting and threshing of rice and different *rabi* crops was done manually to minimize yield losses and reducing experimental error by using sickle and hand operated thresher/by beating the harvested crops with wooden stick.

Table 3.5: Schedule of different cultural operation during *kharif* and *rabi* season

	Cultural operation	Rice	Wheat	Lentil	Castor	Sunflower	Mustard	Onion	Coriander	Fenugreek
	1	2	3	4	5	6	7	8	9	10
1	**Nursery management**									
I	Preparation of nursery bed and seeding	27.06.09	–	–	–	–	–	23.10.09	–	–
II	Manuring and sowing	27.06.09	–	–	–	–	–	23.10.09	–	–
III	Plant protection measure	–	–	–	–	–	–	23.10.09	–	–
IV	Uprooting of seedling	23.07.09	–	–	–	–	–	23.11.09	–	–
2.	**Field preparation**									
I	Ploughing/ Pudding/ Leveling	21.07.09	18.11.09	18.11.09	18.11.09	18.11.09	18.11.09	23.11.09	18.11.09	18.11.09

Continued

	1	2	3	4	5	6	7	8	9	10
3.	**Sowing/ Transplanting, basal application of fertilizer**	23.07.09	20.11.09	20.11.09	20.11.09	20.11.09	20.11.09	23.11.09	20.11.09	20.11.09
4.	**Weeding/ Earthing/ Hoeing**	30.08.09	–	16.01.10 26.12.09	17.1.10 23.12.09	17.01.10 24.12.09	18.01.10 22.12.09	19.01.10 21.02.09	20.01.10 25.12.09	20.01.10 26.12.09
5.	**Irrigations**	3	4	3	2	3	3	8	8	4
6.	**Split dose of Nitrogen**	20.08.09 20.09.09	13.12.09 01.01.10	13.12.09 01.01.10	13.12.09 01.01.10	13.12.09 01.01.10	13.12.09 01.01.10	13.12.09 01.01.10	13.12.09 01.01.10	13.12.09 01.01.10
7.	**Plant protection measures**	21.09.09	–	–	–	–	23.01.10	–	–	–
8.	**Harvesting/ Digging/ Cutting/**	01.11.09	09.03.10	05.03.10	28.03.10	26.03.10	15.03.10	25.03.10	23.03.10 07.01.10 (Cutting)	16.03.10
9.	**Threshing Winnowing**	11.11.09	17.03.10	08.03.10	30.03.10	01.03.10	26.03.10	27.03.10	31.03.10	17.03.10
10.	**Cleaning and bagging**	14.11.09	19.03.10	10.03.10	01.04.10	03.04.10	28.03.10	30.03.10	01.04.10	18.03.10

3.13 Studies on crops

3.13.1 Pre-harvest observation

3.13.1.1 Plant population (No. m^{-2})

Plant population of rice was recorded at 30 DAT and at harvest in one metre square area from five places in each plot. The average plants $^{-2}$ was worked out and expressed in numbers.

3.13.1.2 Plant height (cm)

The plant height of the rice was recorded at 30, 60 DAT and at harvest from five randomly selected plants in each plot and then average was worked out.

3.13.1.3 Dry matter accumulation (g plant ^{1})

Three plants of rice from third row of each plot were randomly selected and uprooted and washed and dried in oven at 105 ^{0}C at 30, 60 DAT and at harvest. Then the samples were weighed on electronic balance and averaged out to get dry matter accumulation in gram plant^{-1}.

3.13.1.4 Leaf area index (LAI)

Three plants uprooted at 30, 60 and 90 DAT for dry matter accumulation study were used for determining the leaf area index. Licor 300 leaf area meter was used to measure the leaf area of uprooted plants and averages were taken to find out the leaf area. LAI was determined with the following formula:-

$$\text{LAI} = \frac{\text{Total leaf area of the hill or plant}}{\text{Total ground area under the hill or plant}}$$

3.14 Weed studies

Studies pertaining to weed dynamics were carried out in rice and following rabi crops during *kharif* and *rabi* season, respectively.

3.14.1 Weed density

Weeds associated with rice and different *rabi* crops in the experimental area were recorded at 30 and 60 DAT/DAS. Total weed count was made in randomly selected three quadrates of 50 cm X 50 cm ($0.25m^2$) from each and every treatment plots. The number of weeds was counted and the data were converted into one square metre area for stastical analysis subjected to square root transformation i.e. $\sqrt{x+1}$.

3.14.2 Dry weight of weeds

Dry weight of weeds was recorded at 30 and 60 DAT/DAS. Weeds present in quadrate of $0.25m^2$ were uprooted along with their roots. The root portion was cut and detached. Only shoot portion of the weed plants were oven dried at 60^0C for 36 hours. After complete oven drying, dry weight of weeds was recorded for different treatment. Dry weight of weeds was also subjected to square root transformation i.e. $\sqrt{x+1}$.

3.15 Post harvest observations

3.15.1 Yield components

3.15.1.1 Panicles plant^{-1}

Number of panicles plant^{-1} of rice was counted from five randomly selected plants and expressed as panicles plant^{-1}.

3.15.1.2 Grains panicle^{-1}

Total number of grains panicle^{-1} of rice was counted from ten panicles and average value was worked out.

3.15.1.3 Panicle length (cm)

The length of panicle was taken form ten panicles selected randomly from harvested produce. It was measured from the neck node to tip of the apical grains. After this, the average length of panicle was determined.

3.15.1.4 Test weight (g)

Five hundred grains were taken from different panicles of each treatment, weighed and multiplied by two to obtain 1000 grain weight. It was expressed in g and converted into 14% moisture in grains.

3.15.2 Biomass production

The crop was harvested from each net plot and then sun dried for two days before taking the total biomass production.

3.15.3 Grain yield (q ha^{-1})

The harvested crop was threshed manually and the grain weight was recorded. Average grain yield of each crop for each plot was worked out separately and converted into q ha^{-1}.

3.15.4 Straw yield (q ha^{-1})

It was calculated by subtracting the grain yield from the total biomass production and expressed in q ha^{-1}.

3.15.5 Harvest index (%)

The harvest index was calculated by using following formula and expressed in percent.

$$\text{Harvest index (\%)} = \frac{\text{Grain yield}}{\text{Total bio mass yield}} \times 100$$

3.15.6 Wheat equivalent yield (kg ha^{-1})

Yield obtained from *kharif* and *rabi* crops was converted into wheat equivalent yield by multiplying yield of respective crop with prevailing farm gate price of produce and divided by price of wheat during 2009-10. prevailing farm gate prices of produce were also used to workout the economics of different cropping systems.

3.16 Chemical analysis

3.16.1 Organic carbon content

Organic carbon content in soil was estimated by Walkey and Black rapid titration method (Piper, 1967).

3.16.2 Available nitrogen

Available nitrogen content in soil after kharif and rabi harvest was determined by alkaline potassium permanganate method described by Subbiah and Asija (1956).

3.16.3 Available phosphorus

Available phosphorus content in soil after *kharif* and *rabi* harvest was extracted by the method described by Olsen (1954).

3.16.4 Available potassium

Available potassium content in soil was extracted by the flame photometer after 5 minutes shaking with 25 ml of 1 N ammonium acetate.

3.17 Economic analysis

Yield of rice and different *rabi* crops was converted into gross return (Rs. ha^{-1}) by multiplying the prevailing market price of individual crop. The cost incurred for using different variable and fixed input was calculated on the basis of prevailing per unit prices of different inputs. Net return was calculated by deducting the cost of cultivation from the gross return. Benefit: cost ratio was obtained by dividing the net return with cost of cultivation. The values of different inputs and cost of cultivation are presented in Appendix – II (a to h). The cost of cultivation and net return of whole system was calculated by simple summing the values of respective crop and subtracting from total gross return.

3.18 System analysis

3.18.1 Productivity efficiency (PE)

Production efficiency was obtained on dividing the total productivity of a system (in terms of rice equivalent yield) by total duration of that system (Tomar and Tiwari, 1990). Total duration of that system includes period required for field preparation and seeding.

Production efficiency (PE) was calculated with the help of following formula:

$$\text{PE (kg ha}^{-1}\text{)} = \frac{\text{Total productivity of system (Eq yield)}}{\text{Total duration of system}}$$

3.18.2 Economic efficiency

The economic efficiency of the systems was calculated on dividing the net returns (Rs ha^{-1}) in a sequence by 365 days and expressed in Rs $ha^{-1}day^{-1}$.

$$\text{Profitability (Rs ha}^{-1}\text{ day}^{-1}\text{)} = \frac{\text{Net return of cropping system}}{365}$$

3.18.3 Relative productivity efficiency (RPE) and relative economic efficiency (REE)

The relative productivity Efficiency (RPE) and relative economic efficiency (REE) were calculated by using following formula (Urkurkar *et al.*, 2007).

$$\text{Relative productivity efficiency (\%)} = \frac{\text{TP of diversified CS-TP of existing CS}}{\text{TP of existing CS}}$$

$$\text{Relative economic efficiency (\%)} = \frac{\text{TP of diversified CS-NR of existing CS}}{\text{NR of existing CS}}$$

where, TP – total productivity, CS – cropping system; NR – net return of the system

The relative productivity efficiency and relative economic efficiency may be negative or positive in terms of percent over existing cropping systems i.e. rice-wheat-fallow.

3.18.4 Irrigation water use efficiency

The irrigation water use efficiency was calculated on dividing the total productivity by total irrigation water used by the crops in that particular sequence.

$$\text{Irrigation water use efficiency (kg ha}^{-1}\text{ cm}^{-1}) = \frac{\text{Grain yield (kg ha}^{-1})}{\text{Irrigation water (cm)}}$$

3.18.4 Employment generation efficiency

Employment generation efficiency was determined dividing the total mans day's employment for the system by 365 days and expressed in percentage.

$$\text{Employment generation efficiency (\%)} = \frac{\text{Man days employment for system}}{365}$$

3.19 Energetics

Energy inputs were calculated and estimated in Mega Joule (MJ) ha^{-1} with reference to the standard values prescribed by Mittal et al. (1985).These inputs were taken to each treatment of rice and rabi crops. Energy values, which were taken for energy estimation. The standard energy coefficient for seed and straw of rice and rabi crops multiplied with their respective yields and summed up to obtained the total energy output. The Energy input for rice, wheat etc. and the cropping system was calculated by adding the respective values under both season crops. Energy output and input ratio, were calculated as per the following formula-

$$\text{Energy output input ratio} = \frac{\text{Energy Output (EO)}}{\text{Energy Input (EI)}}$$

3.20 Statistical analysis

The statistical analysis of data collected on different parameters of rice and rabi crops in different dates was carried out in randomized block design as described by Gomez and Gomez (1984). Test of significance of the treatments was computed with the critical difference at 5 percent probability.

RESULTS AND DISCUSSION

CHAPTER IV

RESULTS AND DISCUSSION

Field experiment on "**Productivity and profitability of intercropping in *rabi* cereal, legume, oilseeds and spices under rice (*Oryza sativa* L.) based cropping system**" was conducted during the *kharif* and *rabi* seasons of 2009-10 at University Research cum Instructional Farm, Indira Gandhi Krishi Vishwavidyalaya, Raipur. Experimental findings on the basis of the various observations on relevant aspects are being presented with Tables and illustrated by suitable Figures and discussed here alongwith the earlier established scientific facts, views and findings of national researchers.

The present study was a part of a long-term experiment of All India Coordinated Research Project on Integrated Farming Systems, which is being conducted during all the three seasons *viz. kharif*, *rabi* and summer with eight different cropping systems since 2009-10 (Table 3.3). As the present study was carried for only *kharif* and *rabi* season crops and the summer component was not taken under the study, two of the treatments seems alike with each other therefore, in spite of presenting the data of all the treatments, one of them *i.e.* T_8 is conceded as dummy of T_1 *i.e.* rice – wheat cropping system and not taken under purview for discussion in this chapter.

4.1 STUDIES IN RICE

Growth characters

4.1.1 Plant population and plant height

As the rice crop was grown under transplanted condition and seedling of rice properly placed in experimental field, the plant population did not influenced significantly due to the different cropping systems and rice varieties grown (Table 4.1). In general, by virtue of vertical growth character of rice plants, their height increased with advancement of crop age. But, the maximum increase was recorded during 30 to 60 DAT, while the pace of the growth was slowed down after 60 DAT (Table 4.2).

Table 4.1: Plant population of three rice varieties as affected by different cropping systems

Tr	Cropping system		Plant population m^{-2}	
	Kharif	Rabi	30 DAT	At harvest
T_1.	Rice MTU 1010	Wheat	48.33	47.67
T_2.	Rice MTU 1010	Castor + Lentil (1:3)	48.33	47.67
T_3.	Rice MTU 1010	Mustard +Lentil (1:2)	48.00	47.33
T_4.	Rice Mahamaya	Sunflower + Lentil (1:3)	48.00	47.67
T_5.	Rice Mahamaya	Wheat +Fenugreek (1:1 skip row)	48.00	48.05
T_6.	Hybrid Rice Indira Sona	Wheat + Lentil (1:1 skip row)	48.00	47.33
T_7.	Hybrid Rice Indira Sona	Onion + Coriander (3:1)	48.00	47.33
T_8.	Rice MTU 1010	Wheat	47.67	47
		SEm±	**0.61**	**0.57**
		CD 5%	**NS**	**NS**

The maximum plant height (66.50 cm) recorded at 30 DAT under T_6 (Hybrid rice 'Indira Sona' - wheat+ lentil) which was significantly more than rice-wheat (60.47 cm), rice-castor +lentil (60.47 cm) and rice-mustard+ lentil (60.50 cm). However, plant height of rice under Hybrid rice-onion+ coriander, rice-sunflower+ lentil and rice-wheat+ fenugreek was similar to that of Hybrid rice-wheat+ lentil. Similar trend was observed at 60 DAT and Hybrid rice

'Indira Sona' - wheat+ lentil again recorded maximum plant height (93.47 cm). The lowest plant height was recorded under rice-wheat (78.17 cm).

However, at harvest owing to the inherent character, all three varieties differ significantly among each other. Hybrid rice Indira Sona showed its superiority and grown higher than other two varieties i.e. MTU-1010 and Mahamaya. The tallest plant at harvest was observed under Hybrid rice-wheat+ lentil (105.07 cm), which was at par with Hybrid rice-onion+ coriander (103.83 cm). The lowest plant height was recorded under rice-castor+ lentil (84.67) and fallowed by rice-wheat (87.67 cm).

Table 4.2: Plant height (cm) of rice varieties as affected by different cropping systems

Tr	Cropping system		Plant height (cm)		
	Kharif	Rabi	30 DAT	60 DAT	At harvest
T_1.	Rice MTU 1010	Wheat	60.47	78.17	88.07
T_2.	Rice MTU 1010	Castor + Lentil (1:3)	60.47	84.10	85.20
T_3.	Rice MTU 1010	Mustard +Lentil (1:2)	60.50	81.70	90.40
T_4.	Rice Mahamaya	Sunflower + Lentil (1:3)	65.43	86.43	101.50
T_5.	Rice Mahamaya	Wheat +Fenugreek (1:1 skip row)	64.50	81.40	100.77
T_6.	Hybrid Rice Indira Sona	Wheat + Lentil (1:1 skip row)	66.50	93.47	105.07
T_7.	Hybrid Rice Indira Sona	Onion + Coriander (3:1)	65.37	92.53	103.83
T_8.	Rice MTU 1010	Wheat	60.53	83.17	88.03
SEm±			1.79	3.06	0.73
CD 5%			5.42	9.27	2.23

4.1.2 Dry matter accumulation g plant $^{-1}$

It is revealed from the Table 4.3 that the dry weight of rice crop was increased with the advancement of growth stage. At 30 DAT, the highest dry matter production (3.88 g plant $^{-1}$) was recorded in Hybrid rice-onion+ coriander (T_7) cropping system followed by Hybrid rice 'Indira Sona' - wheat+ lentil and rice-mustard+ lentil (T_3) cropping system (3.80 g plant^{-1}). Whereas the lowest dry matter production was found in rice-Sunflower +lentil (T_4) cropping system (3.56 g plant^{-1}) followed by rice-castor + lentil (T_2) cropping system (3.60 g plant $^{-1}$).

Table 4.3: Dry matter accumulation g plant $^{-1}$ of three rice verities as affected by different cropping systems.

Tr	Cropping system		Dry weight plant $^{-1}$		
	Kharif	Rabi	30 DAT	60 DAT	At harvest
T_1.	Rice MTU 1010	Wheat	3.72	8.28	28.4
T_2.	Rice MTU 1010	Castor + Lentil (1:3)	3.60	8.08	28.72
T_3.	Rice MTU 1010	Mustard +Lentil (1:2)	3.80	8.12	28.56
T_4.	Rice Mahamaya	Sunflower + Lentil (1:3)	3.56	8.84	30.48
T_5.	Rice Mahamaya	Wheat +Fenugreek (1:1 skip row)	3.68	8.72	30.52
T_6.	Hybrid Rice Indira Sona	Wheat + Lentil (1:1 skip row)	3.80	9.24	30.72
T_7.	Hybrid Rice Indira Sona	Onion + Coriander (3:1)	3.88	9.44	30.64
T_8.	Rice MTU 1010	Wheat	3.60	8.20	28.48
SEm±			0.08	0.08	0.46
CD 5%			0.24	0.24	1.38

With the advancement of crop age, dry matter production differed significantly due to different rice varieties and within the group, treatments performed equally at 60 DAT and at harvest. It is evident that the dry matter production (9.44 g $plant^{-1}$) recorded in hybrid rice 'Indira Sona' – onion + coriander (T_7) cropping system was at par with Hybrid rice 'Indira Sona' – wheat + lentil (T_6) cropping system (9.24 g $plant^{-1}$). Whereas, plant dry weight of Mahamaya under rice – sunflower + lentil (8.84 g $plant^{-1}$) and rice - wheat + fenugreek (8.72 g $plant^{-1}$) and of MTU-1010 under rice – wheat (8.28), rice – castor + lentil (8.08) and rice – mustard + lentil (8.12) was similar. Further, all the three rice varieties differed significantly with each other.

However, at harvest, cropping systems with Mahamaya rice (T_4 and T_5) has produced as much as dry matter (30.48 and 30.52 g $plant^{-1}$) to that of Hybrid rice (30.72 and 30.64 under T_6 and T_7, respectively) and both recorded significantly higher dry matter as compared to those cropping systems where early duration MTU-1010 was grown.

4.1.3 Leaf area index

It is evident that, irrespective of the treatments, leaf area index was increased with the advancement of crop age. At 30 DAT, the highest leaf area index (2.52) was found in T_6 (hybrid rice 'Indira Sona' – wheat + lentil) cropping system followed by T_3 (rice 'MTU 1010' – mustard + lentil) cropping system (2.50). Whereas, the lowest leaf area index was found in T_1 (rice 'MTU 1010' - wheat) cropping system (2.39).

At 60 DAT, the highest leaf area index (5.41 g $plant^{-1}$) was found in T_4 (rice 'Mahamaya' – sunflower + lentil) cropping system followed by T_6

(hybrid rice 'Indira Sona' – wheat + lentil) cropping system (5.34 g $plant^{-1}$). Whereas the lowest leaf area index was found in T_1 (rice 'MTU 1010' - wheat), the existing cropping system (5.17 g $plant^{-1}$) followed by T_2 (rice 'MTU 1010' – castor + lentil) cropping system (5.20 g $plant^{-1}$).

Table 4.4: Leaf Area Index (LAI) of three rice varieties as affected by different cropping systems

Tr	Cropping system		30 DAT	60 DAT	90 DAT
	Kharif	Rabi			
T_1.	Rice MTU 1010	Wheat	2.39	5.17	5.59
T_2.	Rice MTU 1010	Castor + Lentil (1:3)	2.48	5.20	5.59
T_3.	Rice MTU 1010	Mustard +Lentil (1:2)	2.51	5.22	5.63
T_4.	Rice Mahamaya	Sunflower + Lentil (1:3)	2.50	5.41	5.89
T_5.	Rice Mahamaya	Wheat +Fenugreek (1:1 skip row)	2.43	5.33	5.84
T_6.	Hybrid Rice Indira Sona	Wheat + Lentil (1:1 skip row)	2.52	5.34	5.86
T_7.	Hybrid Rice Indira Sona	Onion + Coriander (3:1)	2.48	5.33	5.84
T_8.	Rice MTU 1010	Wheat	2.37	5.20	5.71
SEm±			**0.02**	**0.03**	**0.01**
CD 5%			**0.05**	**0.09**	**0.03**

At 90 DAT, the highest leaf area index was found in T_4 (rice 'Mahamaya' – sunflower + lentil) cropping system (5.89 g $plant^{-1}$) followed by T_6 (hybrid rice 'Indira Sona'- wheat + lentil) cropping system (5.86 g $plant^{-1}$). Whereas, the lowest leaf area index was found in T_1 (rice-wheat) and T_5

(rice- wheat +fenugreek) cropping system (5.59 g plant^{-1}) followed by T_3 (Rice-mustard+ lentil) cropping system (5.63 g plant^{-1}). Magnitude of increase in LAI from 30 DAT to 60 DAT was greater as compared to later stage owing to the maximum tillering occurred at this stage which boosted the dry matter production by utilizing the solar energy and eventually the LAI.

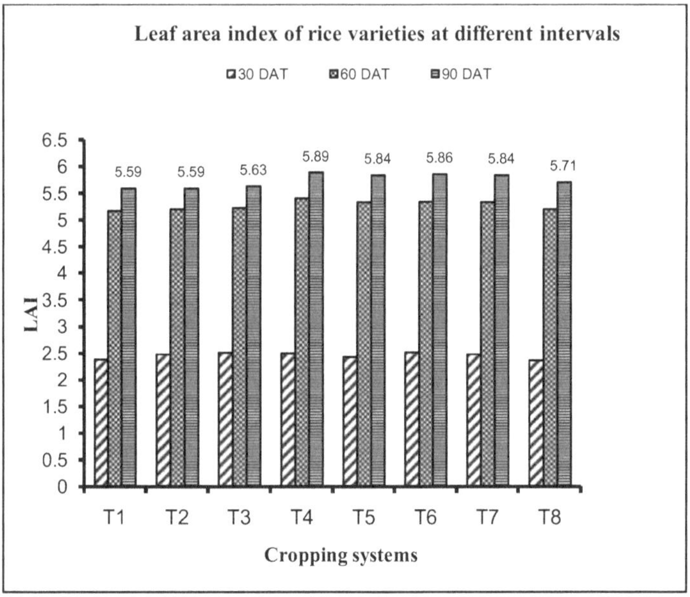

T1: Rice 'MTU-1010'- wheat (existing CS), T2: Rice 'MTU-1010'-castor+lentil, T3:Rice 'MTU-1010'-mustard+lentil, T4: Rice 'Mahamaya'-sunflower+lentil, T5: Rice 'Mahamaya'- wheat+fenugreek, T6: Hybrid rice 'Indira Sona'- wheat+lentil, T7: Hybrid rice 'Indira Sona'-onion+coriander and T8: Rice 'MTU-1010'- wheat

Fig. 4.1 Leaf area index of rice as affected by different cropping systems

4.1.4 Yield attributing characters

Number of panicles plant $^{-1}$

It is clear from the data showed in Table 4.5 that highest number of panicle plant $^{-1}$ (9) was recorded in medium duration rice 'Mahamaya' in T_4 and T_5 and with hybrid rice 'Indira Sona' (9), whereas MTU-1010 produced the lowest number of panicles (7), which is a early duration variety. The highest length of panicle (31.67 cm) was recorded in rice (early duration) T_3 in T_3 Rice-mustard+ lentil cropping system and Hybrid rice (T_7) (31.67 cm) in Hybrid rice-onion+ coriander cropping system.

Table 4.5: Number of panicles, panicle length, grains panicle^{-1} and test weight of three rice varieties as influenced by different cropping system

Tr	Cropping system		Panicles plant^{-1}	Length of panicle (cm)	No. of grains panicle^{-1}	Test weight (g)
	Kharif	*Rabi*				
T_1	Rice MTU 1010	Wheat	7	31.3	92.3	28.00
T_2	Rice MTU 1010	Castor + Lentil (1:3)	7	31.6	96.0	28.67
T_3	Rice MTU 1010	Mustard +Lentil (1:2)	7	29.7	96.0	29.00
T_4	Rice Mahamaya	Sunflower + Lentil (1:3)	9	28.3	106.3	29.00
T_5	Rice Mahamaya	Wheat +Fenugreek (1:1 skip row)	9	29.0	115.3	28.67
T_6	Hybrid Rice Indira Sona	Wheat + Lentil (1:1 skip row)	9	27.3	106.3	30.00
T_7	Hybrid Rice Indira Sona	Onion + Coriander (3:1)	8	31.6	96.0	30.00
T_8	Rice MTU 1010	Wheat	7	30.0	88.3	28.00
SEm±			**0.3**	**2.2**	**6.7**	**0.77**
CD 5%			**1**	**6.7**	**20.3**	**NS**

The lowest length of panicle was recorded in Hybrid rice T_6 (27.33 cm) in Hybrid rice 'Indira Sona' – wheat + lentil, followed by rice 'Mahamaya a medium duration rice in rice-sunflower+ lentil cropping system.

Table 4.5 also describes the results of grains panicle $^{-1}$. The maximum number of grains panicles $^{-1}$ (115.3) was recorded in rice 'Mahamaya' - wheat + fenugreek cropping system, followed by rice 'Mahamaya' – sunflower + lentil and hybrid rice 'Indira Sona'- wheat + lentil cropping system (106.3). The lowest number of grains panicles $^{-1}$ was recorded in earlier duration rice 'MTU-1010 (T_1) (92.3 g) in rice- wheat cropping system.

Although, the maximum test weight of rice (30.0 g) was recorded in Hybrid rice 'Indira Sona'- wheat + lentil (T_6) and hybrid rice 'Indira Sona' – onion + coriander (T_7) cropping systems it was not affected due to cropping systems. The lowest test weight (28.0 g) was recorded in rice 'MTU-1010 in rice- wheat cropping system, followed by rice 'MTU-1010 (28.67 g) in rice – castor + lentil and rice 'Mahamaya' (28.67 g) in rice- wheat + fenugreek cropping system.

It is obvious that yield is the function of over all growth of the plant and various yield attributing components, but it is also important that all these components and growth are attained from how much dry matter is accumulated by the plant/crop earlier. Dry matter is a precursor of how much yield will be achieved at harvest. A handsome part of dry matter accumulated ought to be transformed into higher number of tillers, panicles, grains plant^{-1}, panicle length and test weight to obtain a good harvest.

4.1.5 Grain and straw yield

It is revealed from the Table 4.6 that the maximum grain yield of rice (49.90 q ha^{-1}) was recorded under T_4 (Rice 'Mahamaya' – sunflower + lentil cropping system). This was significantly superior over those cropping systems where early duration rice grown i.e. T_1 (rice 'MTU-1010' - wheat) (41.22 q ha^{-1}), T_2 (rice 'MTU-1010' – castor + lentil) (42.75 q ha^{-1}) and T_3 (rice 'MTU-1010' – mustard + lentil) (43.32 q ha^{-1}). However, medium duration rice produced similar grain yield to that of hybrid rice T_6 (rice 'Indira Sona' – wheat + lentil) (49.70 q ha^{-1}) and T_7 (hybrid rice 'Indira Sona' –onion + coriander) (48.82 q ha^{-1}) and remained at par with T_5 (rice 'Mahamaya' - wheat + fenugreek cropping system (48.68 q ha^{-1}). The rice yield was comparatively low in rice – wheat sequence. The nutrient exploitative nature of both cereals might be the reason for decrease in rice yield (Singh and Sharna, 2002). No significant difference in grain yield of a rice variety among different cropping systems was also reported by Padhi (1993).

Table 4.6: Grain yield, straw yield, and harvesting index of three rice Varieties as affected by different cropping systems

Tr	Cropping system		Grain yield (q ha^{-1})	Straw yield (q ha^{-1})	Harvest index
	Kharif	Rabi			
T_1.	Rice MTU 1010	Wheat	41.22	57.62	41.68
T_2.	Rice MTU 1010	Castor + Lentil (1:3)	42.75	60.56	41.39
T_3.	Rice MTU 1010	Mustard +Lentil (1:2)	43.32	60.38	41.77
T_4.	Rice Mahamaya	Sunflower + Lentil (1:3)	49.90	67.09	42.65
T_5.	Rice Mahamaya	Wheat +Fenugreek (1:1 skip row)	48.68	63.47	43.42
T_6.	Hybrid Rice Indira Sona	Wheat + Lentil (1:1 skip row)	49.70	65.07	43.34
T_7.	Hybrid Rice Indira Sona	Onion + Coriander (3:1)	48.82	63.69	43.39
T_8.	Rice MTU 1010	Wheat	42.63	60.08	41.51
SEm±			**0.94**	**1.32**	**0.61**
CD 5%			**2.85**	**4.01**	**NS**

Straw yield also followed same trend and the maximum straw yield (67.09 q ha^{-1}) was recorded in T_4 (Rice 'Mahamaya' – sunflower + lentil cropping system). This was significantly superior over early duration rice i.e. MTU-1010 under T_1 (57.62 q ha^{-1}, lowest one), T_3 (60.38 q ha^{-1}) and T_2 (60.56 q ha^{-1}). Whereas hybrid rice 'Indira Sona' (T_6 and T_7) remained at par with medium duration rice. Harvest index did not differ significantly among all the three rice varieties (Table 4.6).

Table 4.7: Total weed population and weed dry weight of weeds as affected by different cropping systems

Tr	Cropping system		Total weed population m^{-2}		Weed dry weight m^{-2}	
	Kharif	***Rabi***	**30 DAT**	**60 DAT**	**30 DAT**	**60 DAT**
T_1	Rice MTU 1010	Wheat	10.67 (113.33)	13.07 (170.33)	2.68 (6.70)	14.67 (214.62)
T_2	Rice MTU 1010	Castor + Lentil (1:3)	10.93 (119)	13.10 (171.33)	2.73 (6.96)	14.88 (220.92)
T_3	Rice MTU 1010	Mustard +Lentil (1:2)	10.82 (116.66)	13.05 (170.00)	2.7 (6.82)	15.04 (225.6)
T_4	Rice Mahamaya	Sunflower + Lentil (1:3)	10.78 (115.67)	13.34 (177.33)	2.7 (6.81)	15.02 (225.2)
T_5	Rice Mahamaya	Wheat +Fenugreek (1:1 skip row)	10.72 (114.30)	13.16 (172.67)	2.69 (6.74)	14.81 (218.71)
T_6	Hybrid Rice Indira Sona	Wheat + Lentil (1:1 skip row)	10.59 (111.70)	13.26 (175.33)	2.66 (6.60)	14.92 (222.03)
T_7	Hybrid Rice Indira Sona	Onion + Coriander (3:1)	10.78 (115.67)	13.11 (171.33)	2.7 (6.80)	14.77 (217.58)
T_8	Rice MTU 1010	Wheat	10.69 (113.67)	13.4 (179.00)	2.69 (6.71)	15.09 (227.33)
SEm±			**0.21**	**0.08**	**0.05**	**0.09**
CD 5%			**NS**	**0.23**	**0.15**	**0.27**

Figures in parenthesis show actual plant population

4.1.6 Weed studies

The data pertaining to total weed population in rice at 30 and 60 DAT under different cropping system are presented in Table 4.7. It is obvious from the Table that the rice 'MTU-1010' - castor + lentil (T_2) was highly (119 m^{-2}) infested by weed at 30 DAT followed by rice 'MTU-1010' – mustard + lentil (116.67 m^{-2}). However, during *kharif* no significant difference was observed among all the rice varieties grown at this stage of crop. At 60 DAT, 'Mahamaya' and 'Indira Sona was comparatively more infested by weeds to that of early maturing MTU-1010. This might have been due to medium duration maturing 'Mahamaya' and 'Indira Sona was at tillering stage and there might be more space available for weed growth. Lower weed infestation in early maturing rice might be due to its higher ability to resist weeds and escaping weed competition and attaining maximum growth at this stage. Table 4.7 also describes the weed dry weight at 30 and 60 DAT and it is obvious from the data that dry weight followed almost similar trend to that of weed population.

4.1.7 Available nutrient status at *kharif* harvest

Organic carbon (OC) content of the soil at the time of kharif harvest was not influenced by different cropping sequences. However, slightly higher OC (0.51 g kg^{-1}) was recorded under those treatments where medium duration rice and Hybrid rice was grown to that of early maturing rice treatments (T_1, T_2 and T_3). Available N content in soil was ranged from 241 to 253 kg ha^{-1} under different rice-based cropping system at the time of rice harvest but there was no significant difference was found among the treatments. However, highest available N content (253 kg ha^{-1}) was observed in T_5 (rice

'Mahamaya'- wheat + fenugreek) cropping sequence followed by T_7 (rice 'Indira Sona – onion + coriander) and T_2 (Rice 'MTU-1010 – castor + lentil) (251 kg ha^{-1}) cropping sequence (251 kg ha^{-1}). Higher content of N under T_5 and T_7 was due to higher fertilizer dose of these treatments and might be due to lesser N uptake by early maturing MTU-1010 rice that caused higher N status in soil.

Available P content was also not affected due to different treatments and found little higher in T_5 (rice 'Mahamaya'- wheat + fenugreek cropping sequence (21.4 kg ha^{-1}) followed by T_2 (rice 'MTU-1010' – castor + lentil cropping sequence (20.6 kg ha^{-1}). Similar trend was observed for available k content in soil.

Table 4.8: Effect of different cropping systems on organic carbon and available NPK content in soil at the time of *kharif* harvest

Tr	Cropping system		OC (g kg^{-1})	Available nutrient content in soil (kg ha^{-1})		
	Kharif	Rabi		N	P	K
T_1.	Rice MTU 1010	Wheat	0.50	247	19.3	266
T_2.	Rice MTU 1010	Castor + Lentil (1:3)	0.50	251	20.6	271
T_3.	Rice MTU 1010	Mustard +Lentil (1:2)	0.50	241	18.9	266
T_4.	Rice Mahamaya	Sunflower + Lentil (1:3)	0.51	245	18.5	267
T_5.	Rice Mahamaya	Wheat +Fenugreek (1:1 skip row)	0.51	253	21.4	273
T_6.	Hybrid Rice Indira Sona	Wheat + Lentil (1:1 skip row)	0.50	249	19.5	267
T_7.	Hybrid Rice Indira Sona	Onion + Coriander (3:1)	0.51	251	19.6	270
T_8.	Rice MTU 1010	Wheat	0.50	244	20.2	260
SEm±			**0.03**	**5.27**	**1.00**	**7.5**
CD 5%			**NS**	**NS**	**NS**	**NS**

4.1.8 Economics

It is revealed from Table 4.8 that the maximum net return (Rs 34,704 ha^{-1}) was obtained from early duration rice 'MTU-1010 in rice – mustard + lentil cropping system. This was significantly superior to hybrid rice in T_7 : hybrid rice 'Indira Sona'- onion + coriander (Rs 28,899 ha^{-1}), T_6: hybrid rice 'Indira Sona'- wheat + lentil cropping sequence (Rs 29,769 ha^{-1}) and medium duration rice T_5 : rice 'Mahamaya'- wheat + fenugreek cropping sequence (Rs 31,399 ha^{-1}) whereas, early duration rice under T_1: rice 'MTU-1010'- wheat (Rs 32,194 ha^{-1}), T_2 : rice 'MTU-1010 – castor + lentil (Rs 34019 ha^{-1}) remained at par with T_3. The lowest net return recorded under Hybrid rice based cropping system was due to highest cost of cultivation (Rs 19,926 ha^{-1}) involving higher price of hybrid seed incurred in T_6 and T_7. The highest net return obtained from MTU-1010 was due to higher farm gate price of that rice and proved to be the most remunerable cropping system. Similarly, the maximum B:C ratio (2.01) was recorded in T_3 (rice 'MTU-1010 –mustard + lentil) followed by T_2 : rice 'MTU-1010' – castor + lentil cropping sequence (1.97). Whereas, the lowest B:C ratio (1.45) was calculated in T_7 : rice 'Indira Sona' onion + coriander) followed by T_6 : rice 'Indira Sona'- wheat+ lentil cropping sequence (1.49).

Table 4.9: Cost of cultivation, net return and benefit: cost ratio of rice as affected by different cropping systems

Tr	Cropping system		Cost of Cultivation (Rs ha^{-1})	Net Return (Rs ha^{-1})	B:C ratio
	Kharif	*Rabi*			
T_1.	Rice MTU 1010	Wheat	17281	32194	1.86
T_2.	Rice MTU 1010	Castor + Lentil (1:3)	17281	34019	1.97
T_3.	Rice MTU 1010	Mustard +Lentil (1:2)	17281	34704	2.01
T_4.	Rice Mahamaya	Sunflower + Lentil (1:3)	17281	32619	1.89
T_5.	Rice Mahamaya	Wheat +Fenugreek (1:1 skip row)	17281	31399	1.81
T_6.	Hybrid Rice Indira Sona	Wheat + Lentil (1:1 skip row)	19926	29769	1.49
T_7.	Hybrid Rice Indira Sona	Onion + Coriander (3:1)	19926	28899	1.45
T_8.	Rice MTU 1010	Wheat	17281	33869	1.96
SEm±			-	**1049**	-
CD 5%			-	**3183**	-

4.2 STUDIES IN *RABI* CROPS

During *rabi* season, eight cropping systems evaluated consisting of different main and intercrops having different growth behavior and parameters, which do not give any meaningful inference and could not provide any comparable indications. Hence, only grain yield in terms of wheat equivalent yield, economics and weed dynamics and soil nutrient status are depicted through Tables and duly discussed in this section.

4.2.1 Grain yield in terms of wheat equivalent yield

During *rabi* season, different rice based cropping systems were tested. Some of the systems include intercrops. As all the crops had different growth behavior, the grain yield (actual and intercrop yield) of each system was converted into wheat equivalent yield (WEY) and presented in Table 4.10 and discussed accordingly.

Maximum WEY (65.55 q ha^{-1}) of *rabi* crops was recorded in T_7 (hybrid rice 'Indira Sona' -onion + coriander system. This was significantly superior over all the treatment. Rice 'Mahamaya'- wheat + fenugreek cropping system was next in order which was also recorded significantly higher WEY (36.58 q ha^{-1}) over T_1 : rice 'MTU-1010'- wheat, the existing cropping system (30.61 q ha^{-1}), T_2 : rice'MTU-1010'-castor + lentil (33.17q ha^{-1}) and T_6 : rice 'Mahamaya' – wheat + lentil (28.18 q ha^{-1}). The WEY was maximum due to the higher yield and price of crop. Systems having intercropping of lentil produced lower WEY because lentil is a poor competitive crop due to very slow growth and low plant height compared with sunflower, wheat and mustard which resulted in low yield under intercropping with these crops. Similar results were also reported by Singh *et al.*, (2009) in lentil based intercropping systems and by Singh and Rana (2006) in Indian mustard – lentil intercropping system.

Table 4.10: Grain yield and wheat equivalent yield of *rabi* crops as influenced by different cropping systems

Tr.	Cropping systems		Grain yield ($q\ ha^{-1}$)		Wheat equivalent yield ($q\ ha^{-1}$)
	Kharif	Rabi	Main crop	Intercrop	
T_1.	Rice MTU 1010	Wheat	30.61	-	30.61
T_2.	Rice MTU 1010	Castor + Lentil (1:3)	4.97	9.11	32.78
T_3.	Rice MTU 1010	Mustard +Lentil (1:2)	13.44	4.48	33.17
T_4.	Rice Mahamaya	Sunflower + Lentil (1:3)	16.64	2.25	33.14
T_5.	Rice Mahamaya	Wheat +Fenugreek (1:1 skip row)	16.70	8.52	36.58
T_6.	Hybrid Rice Indira Sona	Wheat + Lentil (1:1 skip row)	19.06	3.80	28.18
T_7.	Hybrid Rice Indira Sona	Onion + Coriander (3:1)	97.77	2.92 (seed) 31.04 (leaf)	65.55
T_8.	Rice MTU 1010	Wheat	28.77	-	28.77
SEm±			-	-	**1.25**
CD 5%			-	-	**3.79**

4.2.2 Weed dynamics

The total weed population and dry matter production of weeds at 30 and 60 DAS is presented in Table 4.11. Perusal of data noticed that significant variation in total weed population and total dry matter production of weeds among different cropping systems. The highest weed population (172.33 and 147.00 m^{-2}) and dry matter production (7.8 and 194 g m^{-2}) was observed in mustard + lentil at 30 and 60 DAS, respectively because of slower growth of mustard and lentil during initial stages and weed competed for light, water and nutrients for quick growth. The lowest weed dry production was found in onion+ coriander (1.8 g m^{-2}) at 30 DAS, and under wheat + fenugreek (82.07 g m^{-2}) at 60 DAS, due to uprooting of weeds during intercultural operation of onion + coriander and faster growth of wheat and fenugreek during earlier stage which did not allow weeds to grow.

Table 4.11: Total weed population and dry matter of *rabi* season weeds as affected by different cropping systems

Tr.	Cropping system		Weed population (m^{-2})		Weed dry weight (m^{-2})	
	Kharif	***Rabi***	**30 DAT**	**60 DAT**	**30 DAT**	**60 DAT**
T_1	Rice MTU-1010	Wheat	11.71 (136.67)	11.45 (130.67)	2.59 (6.2)	12.85 (164.6)
T_2	Rice MTU 1010	Castor + Lentil (1:3)	12.67 (160)	11.91 (141.33)	2.77 (7.2)	12.99 (168.2)
T_3	Rice MTU-1010	Mustard +Lentil (1:2)	13.15 (172.33)	12.14 (147)	2.88 (7.8)	13.95 (194)
T_4	Rice Mahamaya	Sunflower + Lentil (1:3)	11.67 (135.67)	11.39 (129.33)	2.61 (6.3)	13.18 (173.3)
T_5	Rice Mahamaya	Wheat +Fenugreek (1:1 skip row)	7.27 (52.33)	8.48 (71.33)	1.70 (2.4)	9.09 (82.07)
T_6	Hybrid Rice Indira Sona	Wheat + Lentil (1:1 skip row)	11.29 (127)	10.98 (120)	2.54 (5.96)	12.65 (159.6)
T_7	Hybrid Rice Indira Sona	Onion + Coriander (3:1)	6.52 (42)	8.53 (72.33)	1.52 (1.8)	9.15 (83.2)
T_8	Rice MTU-1010	Wheat	11.41 (129.67)	11.50 (131.67)	2.57 (6.1)	12.59 (158)
SEm±			**0.09**	**0.05**	**0.02**	**0.06**
CD 5%			**0.28**	**0.17**	**0.06**	**0.18**

Figures in parenthesis show actual plant population

4.2.3 Available nutrient status at *rabi* harvest

It is clear from the Table 4.12 that the fertility status in soil was influenced by various treatments after completion of one cycle of cropping system. Available organic carbon ranged from 0.50 - 0.54 g kg^{-1}. The organic carbon content was found higher (0.54 g kg^{-1}) in hybrid rice 'Indira Sona'-onion + coriander cropping systems and lower (0.50 g kg^{-1}) in rice 'MTU-1010'-castor + lentil system.

Available N in soil ranged from 252 – 263 kg ha^{-1} under rice based cropping systems and was determined higher in T_7 : hybrid rice 'Indira Sona'-onion + coriander (263 kg ha^{-1}) followed by T_1 :rice 'MTU-1010'- wheat (261 kg ha^{-1}). Whereas, the lowest N content (252 kg ha^{-1}) was recorded in T_2: rice-castor + lentil cropping sequence. Similar trend was observed in case of available P content. Available K content was not influenced by different rice based cropping systems. However, higher K (283 kg ha^{-1}) was determined under T_6 : hybrid rice 'Indira Sona'-wheat + lentil cropping sequence and lower in T_4 :Rice-sunflower+ lentil) (268 kg ha^{-1}).

The reduction in soil fertility due to inclusion of cereal and oilseed cropping sequence may be due to the higher uptake and lower addition of nutrient in soil. The observations were in agreement with Singh *et al.*(2004). The available N, P, K and organic carbon content were increased in soil after the harvest of the *rabi* crops. This might be due to inclusion of pulse crop in rice based cropping systems owing to addition of nutrient by biological N-fixation by these crops. Increase in available N, P, K and OC content in cropping sequences involving vegetable pea, green gram were reported by Gangwar and Ram (2005).

Table 4.12: Available N, P and K and organic carbon (OC) content of soil after *rabi* harvest as affected by different cropping systems

Tr	Cropping systems		OC (g kg^{-1})	Available nutrient content (kg ha^{-1})		
	Kharif	*Rabi*		N	P	K
T_1	Rice MTU 1010	Wheat	0.51	261	20.8	274
T_2	Rice MTU 1010	Castor + Lentil (1:3)	0.52	252	20.8	274
T_3	Rice MTU 1010	Mustard +Lentil (1:2)	0.51	256	20.4	270
T_4	Rice Mahamaya	Sunflower + Lentil (1:3)	0.52	255	20.4	268
T_5	Rice Mahamaya	Wheat +Fenugreek (1:1 skip row)	0.52	257	21.1	274
T_6	Hybrid Rice Indira Sona	Wheat + Lentil (1:1 skip row)	0.53	259	20.4	283
T_7	Hybrid Rice Indira Sona	Onion + Coriander (3:1)	0.53	263	21.6	270
T_8	Rice MTU 1010	Wheat	0.52	257	20.2	264
SEm±			**0.01**	**5.27**	**1.03**	**6**
CD 5%			**NS**	**NS**	**NS**	**19**

4.2.4 Economics

The cost of cultivation, net return and B:C ratio of *rabi* crops are presented in Table 4.13. The economic analysis showed that highest cost of cultivation (Rs.33,033 ha^{-1}) incurred in T_7 : hybrid rice 'Indira Sona' –onion + coriander cropping system as this system required heavy fertilization and irrigation. Cost of seed in onion also increased the cost. The system involving hybrid rice 'Indira Sona' – wheat + lentil and rice 'Mahamaya' – wheat + fenugreek also hiked the cultivation cost considerably (Rs.19,568 and Rs.19,386 ha^{-1}, respectively) because of higher labour requirement for weeding etc. The lowest cost of cultivation (Rs.15,883 ha^{-1}) was calculated in rice 'MTU-1010'-wheat, the existing cropping system.

The highest net returns (Rs.65,292 ha^{-1}) from *rabi* crops were recorded with T_7 : hybrid rice 'Indira Sona'- onion + coriander because of higher value of the produce. It was proved to be the most remunerative cropping system and gave almost double profit and remained significantly superior over all the treatments. Rice 'Mahamaya' – wheat + fenugreek gave considerable net returns (Rs.35,484 ha^{-1}). Whereas, owing to lower grain yield, hybrid rice 'Indira Sona'- wheat + lentil system produced the lowest net return (Rs.22,707 ha^{-1}) among *rabi* crops. The highest B:C ratio of rabi crops was recorded in hybrid rice 'Indira Sona'-onion + coriander (1.98) followed by rice 'MTU-1010'- mustard + lentil (1.90) and rice 'MTU-1010' - wheat (1.89). The lowest B:C ratio of rabi crops (1.16) was obtained in hybrid rice 'Indira Sona'-wheat + lentil.

Table 4.13: Cost of cultivation, net return and B:C ratio of rabi crops as influenced by different cropping systems

Tr	Cropping systems		Cost of cultivation (Rs ha^{-1})	Net return (Rs ha^{-1})	B:C ratio
	Kharif	***Kharif***			
T_1	Rice MTU 1010	Wheat	15,883	30,032	1.89
T_2	Rice MTU 1010	Castor + Lentil (1:3)	18,126	31,049	1.71
T_3	Rice MTU 1010	Mustard +Lentil (1:2)	17,113	32,637	1.90
T_4	Rice Mahamaya	Sunflower + Lentil (1:3)	18,731	30,984	1.65
T_5	Rice Mahamaya	Wheat +Fenugreek (1:1 skip row)	19,386	35,484	1.83
T_6	Hybrid Rice Indira Sona	Wheat + Lentil (1:1 skip row)	19,568	22,707	1.16
T_7	Hybrid Rice Indira Sona	Onion + Coriander (3:1)	33,033	65,292	1.98
T_8	Rice MTU 1010	Wheat	15,883	27,272	1.72
SEm±			-	**1,876**	-
CD 5%			-	**5,690**	-

4.3 TOTAL PRODUCTIVITY AND SYSTEM ANALYSIS

4.3.1 Total productivity in terms of WEY

To have a better understanding of yield data, the yield of rice and *rabi* crops were converted into wheat equivalent yield (WEY) of respective crops and the total productivity of both the seasons in terms of system productivity are presented here in terms of WEY. The winter crops mostly governed the WEY of the system, because rice was the base crop. Maximum total productivity of the system (98.10 q ha^{-1}) in terms of WEY recorded in $T_{7:}$ rice 'Indira Sona – onion + coriander cropping sequence. This system was significantly superior over all the treatments and produced almost 150 % or more yield over other systems. Yield advantage of rice based vegetable cropping systems over rice based system with field crop was also reported by Kumar *et al.*, (2008) and Kalyan Singh *et al.*, (2007).

Further, T_5: rice 'Mahamaya'- wheat + fenugreek system was next in order and produced significantly higher total productivity (69.03 q ha^{-1}) over T_1: rice 'MTU-1010'- wheat (63.59 q ha-) and T_6: rice 'Indira Sona'- wheat + lentil (61.31 q ha^{-1}) and was remained at par with T_4: rice 'Mahamaya'- wheat + lentil (66.41q ha^{-1}), T_2 . rice 'MTU-1010'- castor + lentil (66.98 q ha^{-1}) and T_3 : rice 'MTU-1010'- mustard+ lentil) (67.82 q ha^{-1}). Saroch *et al.* (2005) also reported more productivity of the system by replacing wheat in rice – wheat system with vegetables. Similarly, Singh *et al.* (2001) found that rice- lentil- maize + fodder cowpea system gave significantly highest rice equivalent yield (Fig. 4.2).

4.3.2 Economics of the system

The economic analysis showed in Table 4.14 that the highest cost of production (Rs.52,959 ha^{-1}) incurred in rice 'Indira Sona'- onion + coriander cropping sequence followed by T_6 : rice 'Indira Sona'-wheat + lentil (Rs.39,494 ha^{-1}) rice. Inclusion of onion in the system increased the cost of cultivation, as it required heavy fertilization and irrigation. Cost of seed of onion also increased production cost. The lowest cost of cultivation, was recorded in T_1 : rice 'MTU-1010' - wheat (Rs.33,164 ha^{-1}) cropping sequence followed by T_3 : rice 'MTU-1010' – mustard + lentil (Rs.34,394 ha^{-1}) sequence.

Considering the prevailing market prices of inputs and farm gate prices of different produces, the net return, benefit: cost ratio and profitability were worked out separately to have a proper interpretation of results (Table 4.14). The maximum net return (Rs.94,191 ha^{-1}) was obtained in T_7 : rice 'Indira Sona'-onion + coriander cropping system. This was significantly superior over all the other cropping systems. Cropping systems comprising brinjal, onion and cowpea as vegetables had higher rice equivalent yield and contributed to higher net returns of the systems (Bastia *et al.*, 2008).

Rice 'MTU-1010' – mustard + lentil cropping system (T_3) also found appreciably remunerative and gave total net return of Rs 67,431 ha^{-1} and was significantly superior over the rice 'Indira Sona'-wheat + lentil (Rs.52,476 ha^{-1}) which was lowest. Whereas T_1 :rice 'MTU-1010'- wheat (Rs. 62,226 ha^{-1}), T_2 : rice 'MTU-1010' – castor + lentil (Rs. 65,068 ha^{-1}), T_4 : rice 'Mahamaya'-sunflower + lentil (Rs.63,603 ha^{-1}) and T_5 : rice 'Mahamaya'- wheat + fenugreek (Rs. 66,883 ha^{-1}) was remained at par with T_3. Devi *et al.* (1997) also studied that cropping systems with legumes offered special advantage to farmer and rice - chickpea – maize + cowpea fodder had highest productivity & net profit and the same may be recommended for sustainable cultivation and may be an alternative to rice – wheat sequence.

Table 4.14 : Total Productivity (TP) of the system in terms of wheat equivalent yield (WEY), total cost of cultivation, total net return and benefit: cost ratio of different rice-based cropping systems

Tr	Cropping system		Kharif		Rabi			TP (q ha^{-1})	Cost of Cultivation (Rs ha^{-1})	Net Return (Rs ha^{-1})	B:C ratio
	Kharif	Rabi	Rice yield (q ha^{-1})	WEY (q ha^{-1})	Rabi yield (q ha^{-1}) Main crop	Rabi yield (q ha^{-1}) Inter crop	WEY (q ha^{-1})				
T_1.	Rice MTU 1010	Wheat	41.22	32.98	30.61	-	30.61	63.59	33164	62226	1.88
T_2.	Rice MTU 1010	Castor + Lentil (1:3)	42.75	34.20	4.97	9.11	32.78	66.98	35407	65068	1.84
T_3.	Rice MTU 1010	Mustard +Lentil (1:2)	43.32	34.66	13.44	4.48	33.17	67.82	34394	67431	1.96
T_4	Rice Mahamaya	Sunflower + Lentil (1:3)	49.90	33.27	16.64	2.25	33.14	66.41	36012	63603	1.76
T_5.	Rice Mahamaya	Wheat +Fenugreek (1:1 skip row)	48.68	32.45	16.70	8.52	36.58	69.03	36667	66883	1.82
T_6	Hybrid Rice Indira Sona	Wheat + Lentil (1:1 skip row)	49.70	33.13	19.06	3.80	28.18	61.31	39494	52476	1.33
T_7	Hybrid Rice Indira Sona	Onion + Coriander (3:1)	48.82	32.55	97.77	31.04 (Leaf) 2.92 (Seed)	65.55	98.10	52959	94191	1.78
T_8.	Rice MTU 1010	Wheat	42.63	34.10	28.77	-	28.77	62.87	33164	61141	1.84
SEm			0.94	0.70	-	-	1.25	1.21	-	1808	-
CD 5%			2.85	2.12	-	-	3.79	3.66	-	5485	-

Table 4.15: Production efficiency, economic efficiency, relative productivity efficiency and relative economic efficiency as affected by different cropping systems

Tr	Cropping system		Production efficiency (kg ha^{-1} day^{-1})	Economic efficiency (kg ha^{-1} day^{-1})	Relative productivity efficiency (%)	Relative economic efficiency (%)
	Kharif	Rabi				
T_1.	Rice MTU 1010	Wheat	24.41	170.48	-	-
T_2.	Rice MTU 1010	Castor + Lentil (1:3)	23.40	178.27	5.33	4.57
T_3.	Rice MTU 1010	Mustard +Lentil (1:2)	25.47	184.49	6.65	8.22
T_4.	Rice Mahamaya	Sunflower + Lentil (1:3)	24.83	174.26	4.43	2.21
T_5.	Rice Mahamaya	Wheat +Fenugreek (1:1 skip row)	26.59	183.24	8.55	7.48
T_6.	Hybrid Rice Indira Sona	Wheat + Lentil (1:1 skip row)	24.25	143.77	(-)3.59	(-)15.67
T_7.	Hybrid Rice Indira Sona	Onion + Coriander (3:1)	35.99	258.06	54.27	51.37
T_8.	Rice MTU 1010	Wheat	24.03	167.51	(-)1.13	(-)1.74
SEm±			0.45		-	-
CD 5%			1.36	15.03	-	-

Though, maximum total productivity and total net return was obtained under T_7 : rice 'Indira Sona'-onion + coriander cropping system, but amongst all the systems; T_3 : rice 'MTU-1010' – mustard + lentil was identified to be distinctly economically viable in terms of benefit:cost ratio (1.96). interestingly, rice 'MTU-1010' – wheat which was considered as existing cropping system, gave noticeable return per Re invested (1.88 B:C ratio) inspite of giving lesser total net return because of lower cost of cultivation incurred. The lowest B:C ratio (1.33) was recorded under T_6: rice 'Indira Sona' -wheat+ lentil cropping system. Moderate cost of cultivation by inclusion of legume in system was also gave highest benefit :cost ratio to Bastia *et al.*, (2008).

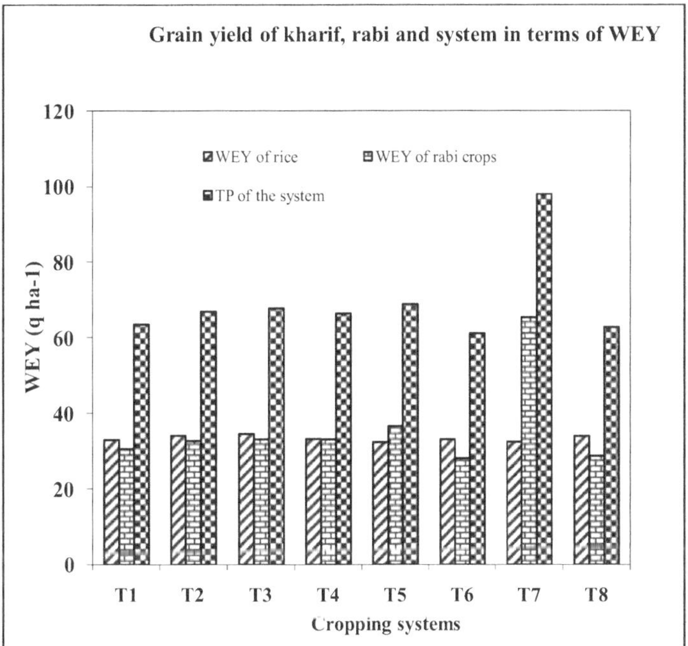

T_1: Rice 'MTU-1010'- wheat (existing CS), T_2: Rice 'MTU-1010'- castor+lentil, T_3:Rice 'MTU-1010'-mustard+lentil,T_4: Rice 'Mahamaya'- sunflower+lentil, T_5: Rice 'Mahamaya'-wheat+fenugreek, T_6: Hybrid rice 'Indira Sona'- wheat+lentil, T_7: Hybrid rice 'Indira Sona'-onion+coriander and T_8: Rice 'MTU-1010'- wheat

Fig. 4.2 : Grain yield of rice, rabi crops and total productivity of the system in terms of wheat equivalent yield (WEY) as affected by different cropping systems

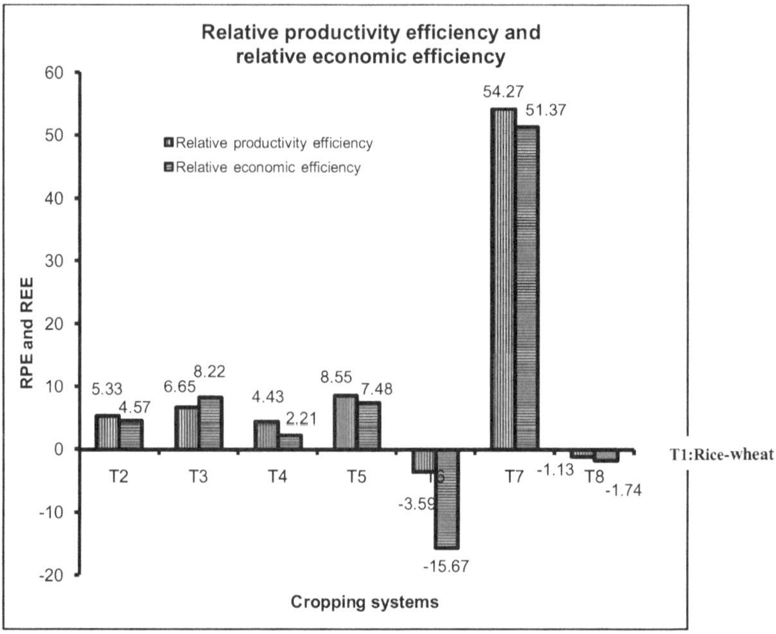

T_1:Rice 'MTU-1010'- wheat (existing CS), T_2: Rice 'MTU-1010'-castor+lentil, T_3: Rice 'MTU-1010'-mustard+lentil,T_4: Rice 'Mahamaya'-sunflower+lentil, T_5: Rice 'Mahamaya'- wheat+fenugreek, T_6: Hybrid rice 'Indira Sona'- wheat+lentil, T_7: Hybrid rice 'Indira Sona'- onion+coriander and T_8: Rice 'MTU-1010'- wheat

Fig. 4.3 : Relative productivity and relative economic efficiency of different cropping systems

4.3.3 Production efficiency and economic efficiency

Crop diversification through crop intensification in time and spatial dimension will not only enhance the productivity of the system but also generates the monetary returns. It is evident from the data presented in Table 4.15 that the production efficiency was highest (35.99 kg $ha^{-1}day^{-1}$) in rice 'Indira Sona'- onion + coriander system which is significantly superior over all other cropping sequence. Production efficiency recorded under $T_{5:}$ rice

'Mahamaya'- wheat + fenugreek was also found significantly higher (26.59 kg ha^{-1} day^{-1}) as compared to T_1, T_2, T_4 and T_6 and at par with T_3 (rice 'MTU-1010' – mustard + lentil). Singh and Singh (2005) also reported that the rice - onion cropping system attained higher production efficiency (33.10 kg ha^{-1} day^{-1}).

The economic efficiency of various rice-base cropping system was calculated and are present in Table 4.15. It was highest in T_7: rice 'Indira Sona'- onion + coriander (Rs 258.06 $ha^{-1}day^{-1}$) and found significantly superior over all other cropping sequence. This might be due to higher value of total produce. Similar results have also reported by Singh *et al.* (1997). The lowest economic efficiency (Rs. 143.77 $ha^{-1}day^{-1}$) was recorded in rice 'Indira Sona'-wheat + lentil cropping sequence.

4.3.4 Relative productivity efficiency and relative economic efficiency

Relative efficiency of different cropping systems have been worked out in respect to total productivity and economics over established rice-wheat cropping system (Table 4.16 & Fig 4.3). It was observed that rice 'Indira Sona'-onion + coriander system had higher positive values in respect of relative productivity efficiency, RPE (54.27%) and relative economic efficiency, REE (51.37%) over rice-wheat cropping system. The rice 'Mahamaya'- wheat + fenugreek, rice 'MTU-1010' – mustard + lentil, rice 'MTU-1010'- castor + lentil and rice - sunflower+ lentil was also found viable in terms of relative productivity and relative economic efficiency point of view. Though, the RPE and REE of rice 'Indira Sona' –wheat + lentil was negative and very less than rice-vegetable system.

4.3.5 Employment generation efficiency

Employment generation efficiency of any diversified system is a direct measure of its preferability in any area as it indicates the employment for farming community for a given period of time. It was evident from the Table 4.17 that employment generation efficiency is ranged between 39.73-75.62%

under different rice based cropping system. Hybrid rice 'Indira Sona'-onion + coriander system coupled with higher monetary returns provided more employment to labourers (276 man days ha^{-1} $year^{-1}$) and thus, its employment generation efficiency was (75.62%) higher as compared to other rice based cropping system. The lowest employment generation efficiency (39.73%) was in T_1 (Rice - wheat) cropping system. The higher labour requirement under T_7 (hybrid rice 'Indira Sona'- onion + coriander) system as the onion crop required intensive labour for various inter cultural operations, viz., hoeing, weeding and fertilizer and irrigation application. Labour employment is a major problem during *rabi* season In Chhattisgarh plain area resulting mass migration of labourers in other state (Urkurkar *et al.*, 2008). Under such conditions the labour intensive system with higher profit can serve the means to provide labour employment.

Table 4.16: Employment generation efficiency, no. of labour employed, irrigation water use efficiency and irrigation water requirement as affected by different cropping systems

Tr	Cropping system		Employment generation efficiency (%)	No. of labour employed	Irrigation water use efficiency (kg ha^{-1} cm^{-1})	Irrigation water requirement (ha-cm)
	Kharif	Rabi				
T_1.	Rice MTU 1010	Wheat	39.73	145	151.41	42
T_2.	Rice MTU 1010	Castor + Lentil (1:3)	53.42	195	223.28	30
T_3.	Rice MTU 1010	Mustard +Lentil (1:2)	55.07	201	188.40	36
T_4.	Rice Mahamaya	Sunflower + Lentil (1:3)	52.60	192	184.47	36
T_5.	Rice Mahamaya	Wheat +Fenugreek (1:1 skip row)	53.97	197	164.36	42
T_6.	Hybrid Rice Indira Sona	Wheat + Lentil (1:1 skip row)	53.97	197	145.99	42
T_7.	Hybrid Rice Indira Sona	Onion + Coriander (3:1)	75.62	276	148.63	66
T_8.	Rice MTU 1010	Wheat	39.73	145	149.69	42
SEm±			-	-	3.20	-
CD 5%			-	-	9.71	-

It is clear from the Table 4.17 That considerable number of labour was also employed under rice MTU-1010'- mustard + lentil (201), rice 'Mahamaya'-wheat+ fenugreek (197) and (hybrid rice 'Indira Sona'-wheat + lentil (197). The lowest employment (145) was available to labourers in rice 'MTU-1010'- wheat) cropping system.

4.3.6 Irrigation water use efficiency and irrigation water requirement

Irrigation water use efficiency was highest with rice 'MTU-1010' – castor + lentil cropping sequence (223.28 kg ha^{-1} cm^{-1}) as less water requirement of castor + lentil intercropping and followed by rice 'MTU-1010'- mustard + lentil (188.40 kg ha^{-1} cm^{-1}) and rice 'Mahamaya'- sunflower + lentil (184.47 kg ha^{-1} cm^{-1}). The lowest irrigation water use efficiency was recorded in Hybrid rice 'Indira Sona'- wheat + lentil (145.99 kg ha^{-1} cm^{-1}). Hybrid rice-onion+ coriander required highest irrigation water (66 ha-cm) due to higher water requirement (almost double to other systems) of onion and coriander at short intervals. The lowest irrigation water (30 ha-cm) was required in rice 'MTU-1010'- castor + lentil. Kumar *et al.*, (2005) also reported that irrigation water requirement of cereal-cereal and cereal-oilseed cropping sequence was found to be lowest as compared to cereal-vegetable sequences due to lesser duration of crop with less number of irrigation required. Growing of cereals with pulses & oilseeds endowed with different root systems help towards better extraction of soil moisture from different layers of the soil, increase water use & water use efficiency as well as intercepting more solar energy and intercropping can give higher yield than when grown as sole crops. (Mandal *et al.*,.1986)

Table 4.17: Total energy input, output, input: output ratio and energy use efficiency of different rice based cropping systems

Tr	Cropping system		Energy Input (MJ x 10^3 ha^{-1})	Energy Output (MJ x 10^3 ha^{-1})	Input / Output ratio
	Kharif	Rabi			
T_1.	Rice MTU 1010	Wheat	20.82	105.62	5.07
T_2.	Rice MTU 1010	Castor + Lentil (1:3)	14.11	88.65	6.28
T_3.	Rice MTU 1010	Mustard +Lentil (1:2)	15.73	98.89	6.29
T_4.	Rice Mahamaya	Sunflower + Lentil (1:3)	15.37	118.26	7.69
T_5.	Rice Mahamaya	Wheat +Fenugreek (1:1 skip row)	17.99	108.61	6.04
T_6.	Hybrid Rice Indira Sona	Wheat + Lentil (1:1 skip row)	18.68	106.66	5.71
T_7.	Hybrid Rice Indira Sona	Onion + Coriander (3:1)	21.41	95.13	4.44
T_8.	Rice MTU 1010	Wheat	20.82	104.95	5.04
SEm±			-	1.46	-
CD 5%			-	4.45	-

4.3.7 Energetics

Table 4.18 depicts the total energy required, total energy output and energy input/output ratio of different rice based systems. Higher energy through fertilizer was required in rice–wheat system and labour requirement for weeding and interculture operations in rice – onion + coriander system. It is evident form the data showed that even consuming highest energy inputs (21.41 MJ x 10^3 ha^{-1}), T_7 : hybrid rice 'Indira Sona' –onion + coriander cropping system failed to generate comparable energy output to the other systems. On the other hand, rice 'Mahamaya'- sunflower+lentil cropping system consumed lower energy (15.37 MJ x 10^3) and produced significantly higher energy output (118.26 MJ x 10^3 ha^{-1}) and gave highest input: output ratio (7.69) over all the treatments. Similar finding was reported by Kumar *et al.* (2005).

SUMMARY, CONCLUSION AND SUGGESTIONS FOR FUTURE WORK

CHAPTER V

SUMMARY, CONCLUSION AND SUGGESTIONS FOR FUTURE RESEARCH WORK

A field experiment entitled **"Productivity and profitability of intercropping in *rabi* cereal, legume, oilseeds and spices under rice (*Oryza sativa* L.) based cropping system"** was carried out at the Research cum Instructional Farm, Indira Gandhi Krishi Vishwavidyalaya (IGKV), Raipur during *kharif* and *rabi* season 2009-10. The soil of the experimental field was silty clay in texture with medium in nitrogen, phosphorus and potassium contents. The experiment was conducted under irrigated conditions because there was least possibility of rains during crop growth period. The treatments consisted of eight (8) cropping sequences *viz* : rice (MTU-1010)-wheat, rice (MTU-1010)- castor + lentil, rice (MTU-1010)- mustard + lentil, rice (MTU-1010)-wheat, rice (Mahamaya)- sunflower+ lentil, rice (Mahamaya)-wheat+ fenugreek, rice (Indira Sona)-wheat+ lentil and rice (Indira Sona)-onion+ coriander were tested in randomized block design (RBD) with three sets in plot size of 87 m^{-2} each. Rice and onion crop was raised by transplanting method. All crops were raised as per the recommended package of practices. The total rainfall received was 1143.3 mm out of which rice crop received 1054.6 mm during the study period.

The salient findings of the present investigation are summarized below:

5.1 Studies on rice

- In general, plant height increased with the advancement of crop age. But, the maximum increase was recorded during 30 to 60 DAT, while the pace of the growth was slowed down after 60 DAT. The maximum plant height

(66.50 cm) recorded at 30 DAT under T_6 (hybrid rice 'Indira Sona' - wheat+ lentil), which was significantly more than rice-wheat (60.47 cm), rice-castor + lentil (60.47 cm) and rice-mustard + lentil (60.50 cm). However, plant height of rice under hybrid rice-onion + coriander, rice-sunflower + lentil and rice-wheat+ fenugreek was similar to that of hybrid rice-wheat + lentil. Similar trend was observed at 60 DAT where hybrid rice 'Indira Sona' – wheat + lentil again produced the maximum plant height (93.47 cm), while the lowest plant height was recorded under rice-wheat (78.17 cm).

- At harvest, hybrid rice Indira Sona showed its superiority and gave taller plants than other two varieties *i.e.* MTU-1010 and Mahamaya. The tallest plant at harvest was observed under hybrid rice-wheat + lentil (105.07 cm), which was at par with Hybrid rice-onion + coriander (103.83 cm).
- The dry matter accumulation of rice crop was increased with the advancement of growth stage. At 30 DAT, the highest dry matter accumulation (3.88 g plant^{-1}) was recorded in hybrid rice-onion+ coriander (T_7) cropping system followed by hybrid rice 'Indira Sona' - wheat+ lentil and rice-mustard + lentil (T_3) cropping system (3.80 g plant^{-1}). At 60 DAT, it is evident that the dry matter accumulation (9.44 g plant^{-1}) recorded in hybrid rice 'Indira Sona' – onion + coriander (T_7) cropping system was at par with hybrid rice 'Indira Sona' – wheat + lentil (T_6) cropping system (9.24 g plant $^{-1}$). Whereas, plant dry weight of Mahamaya under rice – sunflower + lentil (8.84 g plant $^{-1}$) and rice - wheat + fenugreek (8.72 g plant $^{-1}$) and of MTU-1010 under rice – wheat (8.28), rice – castor + lentil (8.08) and rice – mustard + lentil (8.12) was similar. However, at harvest, cropping systems with Mahamaya rice (T_4 and T_5) produced as much as dry matter (30.48 and 30.52 g plant $^{-1}$) to that

of hybrid rice (30.72 and 30.64 under T_6 and T_7, respectively) and both recorded significantly higher dry matter as compared to those cropping systems where early duration MTU-1010 was grown.

- Leaf area index was increased with the advancement of crop age. At 30 DAT, the highest leaf area index (2.52) was found in T_6 (Hybrid rice 'Indira Sona' – wheat + lentil) cropping system followed by T_3 (Rice 'MTU 1010' – mustard + lentil) cropping system (2.50). At 60 DAT, the highest leaf area index (5.41 g plant^{-1}) was found in T_4 (rice 'Mahamaya' – sunflower + lentil) cropping system followed by T_6 (Hybrid rice 'Indira Sona' – wheat + lentil) cropping system (5.34 g plant^{-1}). At 90 DAT, the highest leaf area index was found in T_4 (Rice 'Mahamaya' – sunflower + lentil) cropping system (5.89 g plant^{-1}) followed by T_6 (Hybrid rice 'Indira Sona'- wheat + lentil) cropping system (5.86 g plant^{-1}).
- The highest number of panicles plant $^{-1}$ (9) was recorded in medium duration rice 'Mahamaya' in T_4 and T_5 and with Hybrid rice 'Indira Sona' (9), whereas MTU-1010 produced the lowest number of panicles (7), which is a early duration variety. The highest length of panicle (31.67 cm) was recorded in rice (early duration) T_3 in T_3 Rice-mustard+ lentil cropping system and Hybrid rice (T_7) (31.67 cm) in Hybrid rice-onion+ coriander cropping system.
- The maximum number of grains panicle^{-1} (115.3) was recorded in rice 'Mahamaya' - wheat + fenugreek cropping system, followed by rice 'Mahamaya' – sunflower + lentil and Hybrid rice 'Indira Sona'- wheat + lentil cropping system (106.3). Test weight of rice was not influenced significantly due to cropping systems.

- The maximum grain yield of rice (49.90 q ha^{-1}) was recorded under T_4 (Rice 'Mahamaya' – sunflower + lentil cropping system). This was significantly superior over those cropping systems where early duration rice was grown *i.e.* T_1 (Rice 'MTU-1010' - wheat) (41.22 q ha^{-1}), T_2 (Rice 'MTU-1010' – castor + lentil) (42.75 q ha^{-1}) and T_3 (Rice 'MTU-1010' – mustard + lentil) (43.32 q ha^{-1}). However, medium duration rice produced similar grain yield to that of hybrid rice T_6 (rice 'Indira Sona' – wheat + lentil) (49.70 q ha^{-1}) and T_7 (rice 'Indira Sona' –onion + coriander) (48.82 q ha^{-1}) and remained at par with T_5 (Rice 'Mahamaya' - wheat + fenugreek cropping system (48.68 q ha^{-1}).
- The rice 'MTU-1010' - castor + lentil (T_2) was highly (119 m^{-2}) infested by weed at 30 DAT followed by rice 'MTU-1010' – mustard + lentil (116.67 m^{-2}). However, during *kharif* no significant difference was observed among all the rice varieties grown at this stage of crop. At 60 DAT, 'Mahamaya' and 'Indira Sona was comparatively more infested by weeds to that of early maturing MTU-1010.
- Organic carbon (OC) content of the soil at the time of *kharif* harvest was not influenced by different cropping sequences. However, slightly higher OC (0.51 g kg^{-1}) was recorded under those treatments where medium duration rice and hybrid rice was grown to that of early maturing rice treatments (T_1, T_2 and T_3). No significant difference was found among the treatments. However, highest available N content (253 kg ha^{-1}) was observed in T_5 (rice 'Mahamaya'- wheat + fenugreek) cropping sequence followed by T_7 (rice 'Indira Sona – onion + coriander) and T_2 (Rice 'MTU-1010 – castor + lentil) (251 kg ha^{-1}) cropping sequence (251 kg ha^{-1}). Available P content was also not affected due to different treatments and

found little higher in T_5 (rice 'Mahamaya'- wheat + fenugreek cropping sequence (21.4 kg ha^{-1}) followed by T_2 (Rice MTU-1010' – castor + lentil cropping sequence (20.6 kg ha^{-1}). Similar trend was observed for available K content in soil.

- The maximum net return (Rs 34,704 ha^{-1}) was obtained from early duration rice 'MTU-1010 in rice – mustard + lentil cropping system. This was significantly superior to hybrid rice in T_7 : hybrid rice 'Indira Sona'- onion + coriander (Rs 28,899 ha^{-1}), T_6: hybrid rice 'Indira Sona'- wheat + lentil cropping sequence (Rs 29,769 ha^{-1}) and medium duration rice T_5 : rice 'Mahamaya'- wheat + fenugreek cropping sequence (Rs 31,399 ha^{-1}). Whereas, early duration rice under T_1: rice 'MTU-1010'- wheat (Rs 32,194 ha^{-1}), T_2 : rice 'MTU-1010 – castor + lentil (Rs 34019 ha^{-1}) remained at par with T_3. The maximum B:C ratio (2.01) was recorded in T_3 (rice 'MTU-1010 –mustard + lentil) followed by T_2 : rice 'MTU-1010' – castor + lentil cropping sequence (1.97).

5.2 Studies in *rabi* crops

- Maximum wheat equivalent yield (65.55 q ha^{-1}) of *rabi* crops was recorded in T_7 (hybrid rice 'Indira Sona' -onion + coriander system. This was significantly superior over all the treatment. Rice 'Mahamaya'- wheat + fenugreek cropping system was next in order which was also recorded significantly higher WEY (36.58 q ha^{-1}) over T_1 : rice 'MTU-1010'- wheat, the existing cropping system (30.61 q ha^{-1}), T_2 : rice'MTU-1010'-castor + lentil (33.17q ha^{-1}) and T_6 : rice 'Mahamaya' – wheat + lentil (28.18 q ha^{-1}).

- The highest weed population (172.33 and 147.00 m^{-2}) and dry matter production (7.8 and 194 g m^{-2}) was observed in mustard + lentil at 30 and 60 DAS, respectively.
- Available organic carbon ranged from 0.50 - 0.54 g kg^{-1}. The organic carbon content was found higher (0.54 g kg^{-1}) in hybrid rice 'Indira Sona'-onion + coriander cropping systems and lower (0.50 g kg^{-1}) in rice 'MTU-1010'-castor + lentil system.
- Available N in soil ranged from 252 – 263 kg ha^{-1} under rice based cropping systems and was determined higher in T_7 : hybrid rice 'Indira Sona'-onion + coriander (263 kg ha^{-1}) followed by T_1 :rice 'MTU-1010'-wheat (261 kg ha^{-1}). Similar trend was observed in case of available P content. Available K content was not influenced by different rice based cropping systems.
- The highest net return (Rs.65,292 ha^{-1}) from *rabi* crops were recorded with T_7 : hybrid rice 'Indira Sona'- onion + coriander because of higher value of the produce. It was proved to be the most remunerative cropping system and gave almost double profit and remained significantly superior over all the treatments. The highest B:C ratio of rabi crops was recorded in hybrid rice 'Indira Sona'-onion + coriander (1.98) followed by rice 'MTU-1010'-mustard + lentil (1.90) and rice 'MTU-1010' - wheat (1.89).

5.3 Total productivity and system analysis

- Maximum total productivity of the system (98.10 q ha^{-1}) in terms of WEY was recorded in T_7: rice 'Indira Sona – onion + coriander cropping sequence. This system was significantly superior over all the treatments and produced almost 150 % or more yield over other systems. Further, T_5: rice 'Mahamaya'- wheat + fenugreek system was next in order and produced significantly higher total productivity (69.03 q ha^{-1}) over T_1: rice 'MTU-1010'- wheat (63.59 q ha-) and T_6: rice 'Indira Sona'- wheat + lentil (61.31 q ha^{-1}) but remained at par with T_4: rice 'Mahamaya'- wheat + lentil (66.41q ha^{-1}), T_2 : rice 'MTU-1010'- castor + lentil (66.98 q ha^{-1}) and T_3 : rice 'MTU-1010'- mustard+ lentil) (67.82 q ha^{-1}).

- The maximum net return (Rs.94,191 ha^{-1}) was obtained in T_7 : rice 'Indira Sona'-onion + coriander cropping system. This was significantly superior over all the other cropping systems. Rice 'MTU-1010' – mustard + lentil cropping system (T_3) also found appreciably remunerative and gave total net return of Rs 67,431 ha^{-1} and was significantly superior over the rice 'Indira Sona'-wheat + lentil (Rs.52,476 ha^{-1}) which was lowest. Whereas T_1 :rice 'MTU-1010'- wheat (Rs. 62,226 ha^{-1}), T_2 : rice 'MTU-1010' – castor + lentil (Rs. 65,068 ha^{-1}), T_4 : rice 'Mahamaya'- sunflower + lentil (Rs.63603 ha^{-1}) and T_5 : rice 'Mahamaya'- wheat + fenugreek (Rs. 66,883 ha^{-1}) remained at par with T_3.
- The maximum total productivity and total net return was obtained under T_7 : rice 'Indira Sona'-onion + coriander cropping system, but amongst all the systems; T_3 : rice 'MTU-1010' – mustard + lentil was identified to be distinctly economically viable in terms of benefit: cost ratio (1.96).
- The production efficiency was highest (35.99 kg $ha^{-1}day^{-1}$) in rice 'Indira Sona'- onion + coriander system which is significantly superior over all other cropping sequence. Production efficiency recorded under $T_{5:}$ rice 'Mahamaya'- wheat + fenugreek was also found significantly higher (26.59 kg ha^{-1} day^{-1}) as compared to T_1, T_2, T_4 and T_6 and at par with T_3 (rice 'MTU-1010' – mustard + lentil).

- It was observed that rice 'Indira Sona'-onion + coriander system had higher positive values in respect of relative productivity efficiency, RPE (54.27%) and relative economic efficiency, REE (51.37%) over rice-wheat cropping system. The rice 'Mahamaya'- wheat + fenugreek, rice 'MTU-1010' – mustard + lentil, rice 'MTU-1010'- castor + lentil and rice - sunflower+ lentil was also found viable in terms of relative productivity and relative economic efficiency point of view.
- Employment generation efficiency was ranged between 39.73-75.62% under different rice based cropping system. Hybrid rice 'Indira Sona'-onion + coriander system coupled with higher monetary returns provided more employment to labourers (276 man days ha^{-1} $year^{-1}$) and thus, its employment generation efficiency was (75.62%) higher as compared to other rice based cropping system.
- The higher labour requirement under T_7 (Hybrid rice 'Indira Sona'- onion + coriander) system as the onion crop required intensive labour for various inter cultural operations, viz., hoeing, weeding and fertilizer and irrigation application.
- The considerable number of labour was also employed under rice MTU-1010'- mustard + lentil (201), rice 'Mahamaya'-wheat+ fenugreek (197) and (rice 'Indira Sona'-wheat + lentil (197). Irrigation water use efficiency was highest with rice 'MTU-1010' – castor + lentil cropping sequence (223.28 kg ha^{-1} cm^{-1}) as less water requirement of castor + lentil intercropping and followed by rice 'MTU-1010'- mustard + lentil (188.40 kg ha^{-1} cm^{-1}) and rice 'Mahamaya'- sunflower + lentil (184.47 kg ha^{-1} cm^{-1}).

- Higher energy through fertilizer was required in rice–wheat system and labour requirement for weeding and interculture operations in rice – onion + coriander system. It is evident form the data that even consuming highest energy inputs (21.41 MJ x 10^3 ha^{-1}), T_7 : hybrid rice 'Indira Sona' –onion + coriander cropping system failed to generate comparable energy output to the other systems. On the other hand, rice 'Mahamaya'- sunflower + lentil cropping system consumed lower energy (15.37 MJ x 10^3) and produced significantly higher energy output (118.26 MJ x 10^3 ha^{-1}) and gave highest input: output ratio (7.69) over all the treatments.

CONCLUSIONS

- The cropping systems followed during the *kharif* and *rabi* season had significant effect on grain yield of rice and rabi crops (including intercrops). Rice 'Mahamaya' – sunflower + lentil cropping system produced the highest grain yield (49.90 q ha^{-1}) of rice. Hybrid rice 'Indira Sona'- onion + coriander recorded significantly highest wheat equivalent yield of *rabi* crops (65.55 q ha^{-1}) as well as total productivity of the system (98.10 q ha^{-1}) in terms of wheat equivalent yield.

- The maximum total net return (Rs.94, 191 ha^{-1}) was obtained in hybrid rice 'Indira Sona'- onion + coriander cropping system. This system was significantly superior over all the treatments and generated almost 150 % or more profit over other systems. However, rice 'MTU-1010' – mustard + lentil was identified to be distinctly economically viable in terms of benefit: cost ratio (1.96).

- The production efficiency (35.99 kg $ha^{-1}day^{-1}$) and economic efficiency (Rs. 258.06 $ha^{-1}day^{-1}$) was distinctly highest under the hybrid rice 'Indira Sona'-onion + coriander cropping system. This system was also found the best in terms of relative productivity efficiency, relative economic efficiency. Hybrid rice 'Indira Sona'-onion + coriander system coupled with higher monetary returns provided more employment to labourers (276 man days ha^{-1} $year^{-1}$) and generated highest employment generation efficiency (75.62%).

- Rice 'Mahamaya'- sunflower+lentil cropping system by consuming lower energy (15.37 MJ x 10^3) produced significantly higher energy output (118.26 MJ x 10^3 ha^{-1}) and gave highest input: output ratio (7.69) over all the treatments.

- Available N and P in soil after *rabi* harvest was determined higher in hybrid rice 'Indira Sona'-onion + coriander (263 kg ha^{-1}).

SUGGESTIONS FOR FUTURE RESEARCH WORK

1. The technology for other than rice based cropping system is needed in high rainfall area. This will meet the demand for entrepreneur farmers.
2. The field crops in soybean based cropping system under heavy soil conditions and vegetable based cropping system for upland under irrigated conditions can be tested.
3. Also the spices, legumes & medicinal plants need to be tested in cropping system mode.

4. Legume crop should be included in the rice-wheat cropping system so that the build up of nutrients like N, and fertility of soil may be possible for longer time.
5. In Chhattisgarh, labour employment is major problem during *rabi* season particularly in industrial area resulted mass movement of labours to other states. Under such conditions the labour intensive system with higher profit can provide labour employment.
6. Vegetable crops are highly remunerative, high yielding and provide maximum profit to the farmer, so that vegetables can enhance the socio-economic condition of farmers.
7. Intercropping may be advantageous, when combined with rice or soybean which will make better use of growth factors, lessen risks and thus produces more yield from the system.

ABSTRACT

The experiment was conducted during the *kharif* and *rabi* seasons of 2009-10 at University Research cum Instructional Farm, IGKV, Raipur to study the "**Productivity and profitability of intercropping in *rabi* cereal, legume, oilseed and spices under rice (*Oryza sativa* L.) based cropping system**". The major objectives were to assess the production potential and economic viability of different *rabi* intercropping under rice based cropping systems and to identify the suitable/ remunerative rice based cropping systems with vegetables and oilseeds intercrops. The soil was silty clay (*Inceptisol)* in texture, locally known as "*Matasi*" and neutral in pH and having medium organic carbon and available nitrogen, medium in phosphorus and potassium.

During *kharif* and *rabi*, eight cropping systems were tested viz. Rice 'MTU 1010'- wheat, rice 'MTU 1010'- castor + lentil (1:3), rice 'MTU-1010'-mustard + lentil (1:2), rice 'Mahamaya'- sunflower + lentil (1:3), rice 'Mahamaya'- wheat + fenugreek (1:1 skip row), hybrid rice 'Indira Sona' - wheat + lentil (1:1 skip row), hybrid rice 'Indira Sona'- onion + coriander (3:1) and rice 'MTU 1010'- wheat (dummy treatment) in randomized block design with three replications. The present study was a part of a long-term experiment of All India Coordinated Research Project on Integrated Farming Systems, which is being conducted during *kharif*, *rabi* and summer. While, the present study was under taken only for *kharif* and *rabi* season crops and the summer component was not taken under the study.

The cropping systems followed during the *kharif* and *rabi* season had significant effect on grain yield of rice and *rabi* crops (including intercrops). Rice 'Mahamaya' – sunflower + lentil cropping system produced the highest grain yield (49.90 q ha^{-1}) of rice. While, hybrid rice 'Indira Sona'- onion + coriander recorded significantly highest wheat equivalent yield of *rabi* crops (65.55 q ha^{-1}) as well as total productivity of the system (98.10 q ha^{-1}) in terms of wheat equivalent yield. Despite of highest cost of cultivation incurred (Rs.52, 959 ha^{-1}), the maximum total net return (Rs.94, 191 ha^{-1}) was obtained in hybrid rice 'Indira Sona'-onion + coriander cropping system. This system was significantly superior over all the treatments and generated almost 150 % or more profit over other systems. However, rice 'MTU-1010' – mustard + lentil was identified to be distinctly economically viable in terms of benefit: cost ratio (1.96). Interestingly, despite, of giving lesser total net return rice 'MTU-1010' – wheat which was considered as existing cropping system, gave noticeable return per Re invested (1.88 B:C ratio) because of lower cost of cultivation incurred.

Production efficiency, economic efficiency and employment generation efficiency of any diversified system is a direct measure of its preferability in any area as these indicate the employment for farming community for a given period of time. The production efficiency (35.99 kg $ha^{-1}day^{-1}$) and economic efficiency (Rs 258.06 $ha^{-1}day^{-1}$) was distinctly highest under the hybrid rice 'Indira Sona'-onion + coriander cropping system. This system was also found the best in terms of relative productivity efficiency, relative economic efficiency.

Further, this system coupled with higher monetary returns provided more employment to labourers (276 man days ha^{-1} $year^{-1}$) and generated highest employment generation efficiency (75.62%). On the other hand, irrigation water use efficiency was highest with rice 'MTU-1010' – castor + lentil cropping sequence (223.28 kg ha^{-1} cm^{-1}) as less water requirement of castor + lentil intercropping. Rice 'Mahamaya'- sunflower + lentil cropping system by consuming lower energy (15.37 MJ x 10^3) produced significantly higher energy output (118.26 MJ x 10^3 ha^{-1}) and gave highest input: output ratio (7.69) over all the treatments.

As regards to status of organic carbon and available N, P and K content in the soil, no significant effect was observed during *kharif*. Whereas, Available N and P in soil after *rabi* harvest was determined higher in hybrid rice 'Indira Sona'- onion + coriander (263 kg ha^{-1}). Available K content was not influenced by different rice based cropping systems. The highest weed population (147.00 m^{-2}) and dry matter production (194 g m^{-2}) was observed in mustard + lentil at 60 DAS, due to slower growth of mustard and lentil during early stage and weed competed for light, water and nutrients and suppressed the crop during later stage.

REFERENCES

BIBLIOGRAPHY

Anderson, R. I. 2005. Are some crops synergistic to following crops? *Agronomy Journal* **97**(1): 7–10.

Bastia, D. K., Garnayak, L. M. and Barik, T. 2008. Diversification of rice (*Oryza sativa*) cropping system for higher productivity, resource-use efficiency and economics. *Indian Journal of Agronomy* **53** (1): 22-26.

Bhagat, R.K. and Dhar, V. 2003. Diversification in traditional rice-wheat cropping system. *Journal of Research, Birsa Agricultural University*. 15(2): 237-240.

Bhan, V. M. and Kumar, S. 1998. *Pestology* 20 (4):25-35.

Bharadwaj, V. and Omanwar, P. K. 1994. Long-term effect of continuous rotational cropping and fertilization on crop yields and soil properties. II. Effect on EC, pH, organic matter and available nutrients of soil. *Journal of the Indian Society of Soil Science* **42**:387-392.

Black, C.A. 1965. Method of Soil Analysis. *Amer. Agron. Inc. Madison*. Wisconsin, USA, pp 131-137.

Chauhan, D. S., Sharma, R. K., Kharub, A. S., Tripathi, S. C. and Chhokar, R.S.2001. Effect of crop intensification on productivity, profitability, energetics and soil fertility in rice (*Orizya sativa* L.) wheat (*Triticum aestivum* L.) Cropping system of north –western plain. *Indian Journal of Agricultural Science* **71** (5): 299-302.

Choudhary, J. B., Thakur, R. C., Bhargava, M. and Sood, R. D. 2001. Production potential and economics of rice (*Oryza sativa*)-based cropping systems on farmers' fields under mid-hill conditions of Himachal Pradish. *Himachal Journal of Agricultural Research* **27**(1, 2): 31–35.

Connar, D. J., Gupta, R. K., Hobbs, P. R. and Sayre, K. D. 2003. Bed planting in rice-wheat system. ***In*:** Addressing Resource Conservation Issues in Rice-Wheat Consortium for the Indo-Gangetic Plains, International Maize and Wheat Improvement Center, New Delhi, India, pp. 103-108.

Devi, C. R. S., Nair, G. K. B., Sulochna, K. K, and Nair, V. R. 1997. Integrated nutrient management in a rice – based crop sequence. *International Rice Research news letter* **22** (1): 35 – 36.

Dhurander, R. L., Khanna, P., Pandey, N. and Tripathi, R. S. 2002. Economics and productivity of early-duration-rice (*Oryza sativa* L.) cropping system under assured irrigation in *Alfisol*. *Annuals of Agricultural Research* **23** (4): 669-673.

Dungrani, R.A., Desai, K.D., Patel, M.M. and Patel, M.P. 2005. Productivity and economics of various crop sequences in coastal ecosystem of Gujarat. *Journal of Farming System Research and Development* **11**(1): 69-71.

Fujita, K., Ofosu-Budu, K.G., Ogata, S. 1992. Biological nitrogen fixation in mixed legume-cereal cropping systems. Plant Soil **141:** 155-175.

Gangwar , B. , Katyal , V. and Anand , K.V. 2004 . Stability and efficiency of cropping systems in Chhattisgarh and Madhya Pradesh. *Indian Journal of Agricultural Sciences* **74** (10): 521 – 528.

Gangwar, B., and Katyal, V. 2001. Productivity, stability and profitability of rice (*Oryza sativa*) -based crop sequences in West Bengal and Orissa. *Indian Journal of Agronomy* **46**(3): 144–148.

Gangwar, B., and Ram, B. 2005. Effect of crop diversification on productivity and profitability of rice (*Oryza sativa*)-wheat (*Triticum aestivum*) system. *Indian Journal of Agricultural Sciences* **75** (7): 435–438.

Ganvir, M.M., Jadhao, P.N., Raut, R.F., Shamkuwar, G.R. and Tagade, U.G., 2004. Studies on castor based intercropping system under dryland conditions. *Annals of Plant Physiology* **18**(1): 55-57.

Ghanbari-Bonjar, A., Lee, H.C. 2002. Intercropped field beans (*Vicia faba*) and wheat (*Triticum aestivum*) for whole crop forage: effect of nitrogen on forage yield and quality. J Agr Sci **138:** 311-315.

Ghosh, A.B. 1987. Some aspects of sustainability in soil fertility and crop production. Journal of Indian Society of Soil Science **35** (4): 552-565.

Gill, H.S. and Brar, L.S. 1985. Proc. Indian Society of Weed Science held at GAU, Anand during April 4-5, 1985, pp 98-99.

Gomez, K.A. and Gomez, A.A. 1984. Statistical procedures for Agricultural Research. A Willey- Interscience Publication, John Willey and Sons, New York.

Horwith, B. 1985. A role for intercropping in modern agriculture. BioScience **35:** 286-291.

Jackson, M.L. 1967. Soil Chemical Analysis. Prentice Hall of India Pvt. Ltd., New Delhi. pp 42-48.

Jat, L.N. and Singh, S.M. 2003. Varietal suitability, productivity and profitability of wheat (*Triticum species*) intercrops and relay cropping under furrow-irrigated raised bed system. *Indian Journal of Agricultural Sciences* **73**(4): 187-190.

Kharub, A.S., Chauhan, D.S., Sharma, R.K., Chhokar, R.S. and Tripathi, S.C. 2003. Diversification of rice (*Oryza sativa* L.) –wheat (*Triticum aestivum*) system for improving soil fertility and productivity. *Indian Journal of Agronomy* **48**(3): 149 – 152.

Kumar, A. and Yadav, D.S. 1993. Effect of long term fertilization on soil fertility and yield under rice-wheat cropping system. *Journal of Indian Society of Soil Science* **41**(1): 178-180.

Kumar, A., Tripathi, H.P., Yadav, R.A. and Yadav, D.S. 2008. Diversification of rice (*Oryza sativa*) - wheat (*Triticum aestivum*) cropping system for sustainable production in eastern Uttar Pradesh. *Indian Journal of Agronomy* **53** (1): 18-21.

Kumar, A., Yadav, D.S., Singh, R.M. and Achal, R. 2001. Productivity, profitability and stability of rice (*Oryza sativa*) – based cropping systems in eastern Uttar Pradesh. *Indian Journal of Agronomy* **46**(4): 573–577.

Kumar, K. Avil, Reddy, N.V. and Sadasiva Rao, K. 2005. Profitable and energy efficient rice-based cropping systems in northern Telangana of Andhra Pradesh. *Indian Journal of Agronomy* **50**(1): 6–9.

Kumpawat, B.S. 2001. Production potential and economics of different crop sequences *Indian Journal of Agronomy* **46**(3):421-424.

Mahapatra, B.S., Kumar, A., Shukla, D.K. and Shukla, R.K. 2002. Summer legume in relation to productivity and fertility in rice-wheat cropping system. Paper presented at the International Agronomy Congress, held during 26-30 Nov. 2002, New Delhi, Vol.- 1 pp 155-156.

Mallik, A., Verma, U.N., Thakur, R. and Srivastava, V.C. 1993. Productivity of wheat (*Triticum aestivum*)-based cropping system under limited irrigation. *Indian Journal of Agronomy* **38**(2): 178-181.

Mandal, B.K., Dasgupta, S. and Ray, P.K. 1986. Yield of wheat, mustard and chickpea grown as sole and intercrop with four moisture regimes. *Indian Journal of Agricultural Sciences* **56** (8): 577 – 583.

Manjunath, B.L. and Korikanthimah, V.S. 2004. Productivity under different rice-based cropping systems and physio-chemical properties of soil as influenced by source of manure in Coastal Eco-system of Goa. *Journal Farming Research and Development* **10** (1 & 2): 33–40.

Mittal, V.K., Mittal, J.P. and Dhawan, K.C. 1985. Resource digestion on energy requirement in agriculture sector (1972-82). ICAR/AICRIP/ERAS/85 (1), Ludhiana pp 159-163.

Nambiar, K.K.M. and Abrol, I.P. 1989. Long term fertilizer experiments in India. An Overveiw *Fertilizer News* **34** (4): 11 – 20.

Nayak, B.C., Mishra, S.S. and Patra, A.K. 2003. Production potential and economic analysis of rice (*Oryza sativa*)-based cropping systems. *Indian Journal of Agricultural Sciences* **73**(10): 556-557.

Odhiambo, G.D., Ariga, E.S. 2001. Effect of intercropping maize and beans on striga incidence and grain yield. Seventh Eastern and Southern Africa Regional Maize Conference, 183-186.

Olsen, S.R. 1954. Estimation of available phosphorus in soils by extraction with sodium bi-carbonate USDA. Circ. 939: 1-19.

Padhi, A.K. 1993. Productivity and economics of rice-based cropping sequences. *Indian Journal of Agronomy* **38**(3): 351–356.

Padmavath,P. and Raghavaiah, C.V. 2004. Productivity and returns of castor (*Ricinus communis*)-based intercropping systems with pulses and vegetables under rainfed conditions. *Indian Journal of Agricultural Sciences* **74**(5): 235-238.

Panwar, K.S. and Garg, R. 2004. Studies on the intercropping of pulses and oilseeds in *Brassica napus* L. *Annals of Agri Bio Research* **9**(2): 193-196.

Papanastasis, V.P., Arianoutsou, M., Lyrintzis, G. 2004. Management of biotic resources in ancient Greece. Proceedings of the 10th Mediterranean Ecosystems (MEDECOS) Conference, 25 April-01 May 2004, Rhodes, Greece, pp. 1-11.

Parihar, S.S. Shukla, R.K., Pandey D.,Verma, V.K. and Sharma, R.B. 1995. Productivity and economics of rice (*Oryza sativa*) –based cropping sequence on clay loam soil. *Indian Journal of Agronomy* **40**(4):535.

Parihar, S.S., Pandey, R.K., Shukla, R.K., Verma, K.V., Choure, N.K., Choudhary, K.K. and Pandey, K.S. 1999. Energetic and water use efficiency and economics of rice-based cropping system. *Indian Journal of Agronomy* **44** (2): 205–209.

Piper, C.S. 1967. Soil and Plant Analysis. Bombay, New Delhi, Asia Publishing House. pp 30-38.

Prasad, K. Rawat, S.K. and Pyare, R. 2003. Performance of mustard varieties with association of chickpea under limited irrigations. *Farm Science Journal* **12**(2): 157-158.

Quayyam, M.A. and Maniruzzaman, A.F.M. 1996. Effect of preceding crops on yield of succeeding transplanted Aman rice (*Oryza sativa* L.). *Indian Journal of Agronomy* **41**(3): 349-353.

Russell, A.E. 2002. Relationship between crop-species diversity and soil characteristics in southwest Indian agroecosystems. Agr Ecosyst Environ **92:** 235-249.

Sarkar, R.K., Malik, G.C. and Pal, P.K. 2004. Effect of intercropping lentil (*Lens culinaris*) and linseed (*Linum usitatissimum*) under varying plant density and row arrangement on productivity and advantages in system under rainfed upland. *Indian Journal of Agronomy* **49**(4): 241-243.

Saroch, K., Bhargava, N. and Sharma, J.J. 2005. Diversification of existing rice (*Oryza sativa* L.)-based cropping system for sustainable productivity under irrigated conditions. *Indian Journal of Agronomy* **50** (2); 86 -88.

Saumi, R.C., Kumar, A.L., Majumdar, D., Mani, P.K. and Sahu, P.K. 2004. Diversification of rice (*Oryza sativa*)-based cropping system in New Alluvial Zone of the West Bengal. *Indian Journal of Agronomy* **49**(2): 71-73.

Setty, T.K.P. and Gowda, N.A.J. 1997. Performance of rice-based cropping systems in coastal Karnataka. *Indian Journal of Agronomy* **42**(1): 5–8.

Sharma, R.P., Pathak, S.K., Haque, M. and Lal, M. 2008. Productivity, profitability and nutrient balance as influenced by diversification of rice (*Oryza sativa*)- wheat (*Triticum aestivum*) cropping system. *Indian Journal of Agronomy* **53** (2): 97-101.

Sharma, R.P., Pathak, S.K., Haque, M. and Raman, K.R. 2004. Diversification of traditional rice (*Oryza sativa*) - based cropping system for sustainable production in south Bihar alluvial plains. *Indian Journal of Agronomy* **49**(4): 118–122.

Sharma, R.P., Pathak, S.K., Jha, R.N., Chattopadhaya, N. and Lal, M. 2006. Diversification of existing rice - wheat cropping system for sustainable productivity and profitability under irrigated conditions. *In: Symposium on Conservation and Management of Agro-resources in*

Accelerating the Food production for 21st Century. Dec. 2006. Indian Society of Agronomy Raipur Chapter Department of Agronomy Indira Gandhi Krishi Vishwavidayalaya, Raipur (C.G.)-492006.

Sharma, S.K. and Sharma, S.N. 2002. Balance sheet of nitrogen, phosphorus and potassium under different rice (*Oryza sativa*) – based cropping systems. *Indian Journal of Agronomy* **47**(1): 6–11.

Singh, A.P. and Tuteja, S.S. 2000. Productive potential and economic assessment of rice based cropping systems under irrigation and rainfed conditions. *Indian Journal of Agricultural Issues* **5** (172): 47-50.

Singh, G., Mehta, R.K., and Singh, R.V. 2001. Energetic, yield, water use and economics of rice (*Oryza sativa*)-based cropping system in flood-prone situation of eastern Uttar Pradesh. *Indian Journal of Agronomy* **46**(4): 578-583.

Singh, G., Mehta, R.K., Kumar, T., Singh, R.G., Singh, O.P. and Kumar, V. 2004. Economics of rice (*Oryza sativa* L.)-based cropping system in semi-deep and flood-prone situation in eastern Uttar Pradesh. *Indian Journal of Agronomy* **42** (1): 9-12.

Singh, N.P. and Singh, S. 1983. Compatibility and water use studies in intercropping system of chickpea with different wheat plant types under limited water supply. *Annals of Agricultural Research* **4** (1-2): 140 – 149.

Singh, R.D., and Prasad, U.K. 1994. Effect of irrigation on nutrient balance in rice (*Oryza sativa*) -based cropping sequences. *Indian Journal of Agronomy* **39** (3): 356-362.

Singh, S.N., Sah, A.K., Singh, R.K., Singh, V.K. and Hasan, S.S. 2010. Diversification of rice (*Oryza sativa* L.)-based crop sequences for higher production potentials and economic returns in India's Central Uttar Pradesh. *Journal of Sustainable Agriculture* **34**:141-152.

Singh, T. and Rana, K.S. 2006. Effect of moisture conservation and fertility on Indian mustard (*Brassica juncea*) and lentil (*Lens culinasis*) intercropping system under rainfed conditions. *Indian Journal of Agronomy* **51**(4): 267-270.

Singh, V. and Singh, S. 2005. Productivity and economic of rice-based cropping sequences on recently reclaimed soil, Annals Plant and Soil Research **7** (2): 154-156.

Singh, V., Sood, A.A., Hasan, B., Singh, P. and Singh, S.R. 2008. Production potential and economics of intercropping of lentil (*Lens culinaus*) with brown sugar (*Brassica compestris*) and Oat (*Avena sativa*). *Indian Journal of Agronomy* **53**(2): 135-139.

Singh, V.K. and Sharma, B.B 2002. Economic evaluation of rice (*Oryza sativa*)-based cropping sequences in the foothills of Himalayas. *Indian Journal of Agronomy* **47** (1): 12-19.

Singh, Y., Chaudhary, D.C., Singh, S.P., Bhardwaj, A.K. and Singh, D. 1996. Sustainability of rice-wheat sequential cropping through introduction of legume crops and green manure crops in the system. *Indian Journal of Agronomy* **41** (4), 510-514.

Soria, J.R., Bazan, A., Fargas, J. 1975. Investigation of intercropping in pest management in corn-cassava and bean- cassava intercropping. Experimental Agriculture, **32:** 283-295.

Srinivasan, G., Pothiraj, P. and Shankaran, S. 1992. *Indian Journal of Agronomy* **37** (1): 13-17.

Subbiah, B.V. and Asija, G.C. 1956. A rapid method for the estimation of nitrogen in soil. *Current Science* 26: 259-288.

Thakur, N.S., Sharma, R.S. and Singh, P. 2004. Study on sunflower (*Helianthus annuus L.*) based intercropping systems for Satpura plateau zone of Madhya Pradesh. *Journal of Oilseeds Research* **21**(1): 192-193.

Thakur, R.B., Jha, G. and Choudhary, S.K. 1998. Production and economics of rice (*Oryza sativa*)-based cropping systems. *Indian Journal of Agronomy* **43** (4): 572-575.

Tiwari, A.N.,Singh V.,Rathi, K.S. and Singh, R.D. 1990. *Indian Journal of Agronomy* **35**(1&2): 210-211.

Tomar, S. and Tiwari, A.S. 1990. Production and economics of different crop sequences. *Indian Journal of Agronomy* **35** (1 & 2): 30-35.

Urkurkar, J.S., Chitale, S, Tiwari, A., Savu, R. M. and Tomar, H.S. 2008. Identification of promising rice (*Oryza sativa*) based cropping system for increasing productivity and sustainability for Chhattisgarh plains. *Farming System Research and Development* **14** (1):50-55.

Urkurkar, J.S., Nain, A.S., Tiwari, A and Mishra, R.K., 2007. Analysis of cropping system in Chhattisgarh. AICRP- Cropping System Research, Development of Agronomy, Directorate of research. pub. by IGKV, Raipur(C.G.)

Willy, R.W., et al. 1983. Intercropping studies with annual crops. In: Better Crops for Food, CIBA Foundation Symposium 97. Pitman, London, UK.

Yadav, D.S., Singh, R.M., Kumar, A., Achal, R. 2005. Diversification of traditional cropping system for sustainable production. *Indian Journal of Agronomy* **45**(1): 37–40.

Yadav, D.S., Singh, S.B. and Achal, R. 2000, Production potential and economic feasibility of different rice-based cropping system under sub-humid conditions of eastern U.P. *Narendra Dev Journal of Agricultural Sciences* **4**(1): 147–150.

APPENDICES

Appendix I: Weekly metrological data during the crop period (July 01, 2009 to March 31, 2010)

Month and year	Standard Week No.	Temperature (^{0}C)		Rainfall (mm)	Relative Humidity (%)		Wind velocity (kmph)	Evaporation (mm)	Sun shine (Hours)
		Max.	Min.		I	II			
July, 2009	27	33.0	25.7	59.7	90	67	8.0	4.8	2.9
	28	29.4	24.8	334.0	91	80	11.1	2.4	0.5
	29	28.7	25.0	227.6	89	81	13.0	3.1	1.3
	30	29.8	25.3	13.0	85	73	12.5	4.1	3.8
	31	32.9	26.6	2.0	81	57	9.2	5.5	6.4
Average		**30.76**	**25.48**	**636.3***	**87.2**	**71.6**	**10.76**	**3.98**	**2.98**
August,2009	32	31.8	25.4	99.6	87	74	8.5	4.9	3.8
	33	30.7	25.4	50.2	93	77	6.1	3.7	3.2
	34	30.2	24.4	99.2	93	78	6.3	3.5	3.1
	35	30.8	25.2	17.2	95	83	4.8	3.3	3.9
Average		**30.88**	**25.10**	**266.2***	**92**	**78**	**6.43**	**3.85**	**3.50**
September,2009	36	31.2	25.3	17.2	95	83	4.8	3.3	3.9
	37	32.2	24.7	8.6	92	62	5.0	4.2	6.6
	38	33.0	25.8	0.4	92	62	3.3	4.3	7.7
	39	32.9	25.1	6.4	90	65	4.3	3.9	4.3
Average		**32.3**	**25.22**	**97.2***	**91**	**66.75**	**4.92**	**4.2**	**5.725**
October,2009	40	30.6	24.8	50.8	92	79	6.5	3.7	4.4
	41	31.9	22.1	0.8	93	93	1.9	3.3	6.9
	42	32.7	20.0	0.0	92	92	41	1.7	3.9
	43	31.3	15.3	0.0	94	25	2.0	4.1	9.3
	44	30.7	16.7	0.0	92	34	1.9	3.1	7.6
Average		**31.4**	**19.7**	**51.6***	**92.6**	**46.4**	**2.8**	**3.6**	**7.4**
November,2009	45	30.4	20.9	0.0	90	56	3.4	3.3	4.6
	46	29.8	21.6	49.5	91	68	4.0	3.7	3.6
	47	26.4	13.0	0.0	93	40	4.8	2.5	7.8
	48	26.9	10.8	0.0	93	32	1.6	2.8	8.9

Month and year	Standard Week No.	Temperature (^{0}C)		Rainfall (mm)	Relative Humidity (%)		Wind velocity (kmph)	Evaporation (mm)	Sun shine (Hours)
		Max.	Min.		I	II			
Average		**28.38**	**16.58**	**49.5***	**91.75**	**49.00**	**3.45**	**3.08**	**6.23**
December,2009	49	28.6	13.1	0.0	93	36	1.5	2.6	7.7
	50	29.5	14.9	0.0	93	39	1.7	2.8	7.1
	51	26.1	13.5	19.2	94	52	2.0	2.5	6.4
	52	26.1	11.6	0.1	91	36	2.0	2.6	7.1
Average		**27.58**	**13.28**	**19.3***	**4.83**	**92.75**	**40.75**	**1.80**	**2.63**
January,2010	01	25.6	08.6	0.0	88	29	2.2	2.9	8.4
	02	26.6	14.7	15.4	87	49	2.0	2.5	5.1
	03	25.7	09.3	0.0	94	32	1.3	2.6	7.6
	04	27.8	08.7	0.0	84	21	1.7	3.3	9.5
	05	28.5	11.1	0.0	85	34	2.3	3.6	8.2
Average		**26.84**	**10.48**	**15.4***	**87.6**	**33**	**1.9**	**2.98**	**7.76**
February,2010	06	28.6	13.2	1.4	86	29	2.0	2.8	7.0
	07	31.7	17.4	0.0	88	43	1.7	4.1	8.6
	08	32.1	15.0	5.6	80	26	3.0	5.2	9.4
	09	34.4	16.7	0.0	70	21	2.8	5.4	10.1
Average		**31.70**	**15.58**	**7***		**81.00**	**29.75**	**2.38**	**4.38**
March,2010	10	36.0	20.2	0.0	63	23	3.8	6.8	9.7
	11	36.3	19.3	0.8	64	24	3.7	6.2	8.1
	12	39.1	20.4	0.0	58	14	3.3	7.6	9.4
	13	40.4	24.8	0.0	55	17	6.9	9.8	8.6
		37.95	21.18	0.8*	60.00	19.50	4.43	7.60	8.95
				1143.3**					

* Total of the month, ** Total of the crop period

Appendix –II a :

Cost of cultivation of rice (Rs. ha^{-1})

SN	Particular	Input	Rate Rs	MTU -1010	Mahamaya	Indira Sona
1.	**Field Preparation**					
	ploughing	4 hrs	300/hr	1200	1200	1200
	harrowing	4 hrs	300/hr	1200	1200	1200
	puddling and leveling	3 hrs	300/hr	900	900	900
	labourer	2	100/head	200	200	200
2.	**Seeding**					
	Seed cost					
	MTU -1010	40 kg	11.5/hg	460		
	Mahamaya	40 kg	11.5/hg		460	
	Indira Sona	20 kg	150/kg			3000
3.	**Nursery management**					
	Field preparation			250	250	250
	labourer	10	100/head	1000	1000	1000
4.	**Fertilizer**					
	MTU 1010, Mahamaya					
	N	100 kg	5.52/kg	552	552	
	P	60 kg	14.38/kg	863	863	
	K	40 kg	3.9/kg	156	156	
	Indira Sona					
	N	120 kg	5.52/kg			662
	P	60 kg	14.38/kg			858
	K	40 kg	3.9/kg			156
5.	**Weeding**					
	first	20	100/head	2000	2000	2000
	second	20	100/head	2000	2000	2000
	Irrigation charge	3	300/irri	900	900	900
	labourer	6	100/head	600	600	600
6.	**Plant protection**	1 spray/2 labor	500/spray	500	500	500
7.	**Harvesting**	20 labourer	100/head	2000	2000	2000
8.	**Threshing/winnowing**	20 labourer	100/head	2000	2000	2000
9.	**Land rent**	one season	500	500	500	500
	Total			**17281**	**17281**	**19926**

Appendix –II b

Cost of cultivation of wheat (Rs. ha^{-1})				
S.N.	**Particulars**	**Input**	**Rate (Rs)**	**Cost (Rs)**
1	Field preparation			
	Ploughing by tractor	4 hrs	400/hr	1600
	Harrowing and leveling	5 hrs	400/hr	2000
	cleaning of field	2 labourer	100/head	200
2	Seeding cost @ 100kg/ha	100 kg	12/kg	1200
3	seed treatment by Bavistin	300 g @ 3g/kg of seed	450/kg	135
	Seeding by ferti-seed drill	2 hrs +2 labrourer	400/hr	1000
4	Fertilizers			
	100:60:40 kg/ha N:P:K	N: 100 kg	12.13/kg	1213
		P_2O5: 60 kg	22.12/kg	1327
		K_2O 40 kg	7.70/kg	308
5	Weeding			
	Post emergance herbicide	2 4 D @ 1.5 kg/ha +1 labr	200/kg	600
6	Irrigation	4 numbers	300/irrigation	1200
		4 labourer	100/head	400
7	Plant protection	One spray	600/spray	600
		1 labour	100/head	100
8	Harvesting	20 labourer	100/head	2000
9	Threshing and winnowing	20 labourer	100/head	2000
10	Land rent for one season		500/ season	500
	Total cost Rs ha^{-1}	54 labourer		**16383**

Appendix –II c

Cost of cultivation (Rs/ha) of wheat+ lentil (1:1) skip/alternate row of 20 cm (50%-50%)				
S.N.	**Particulars**	**Input**	**Rate (Rs)**	**Cost (Rs)**
A	**Common or Fixed cost**			
1	Field preparation			
	Ploughing by tractor	4 hrs	400/hr	1600
	Harrowing and leveling	5 hrs	400/hr	2000
	cleaning of field	2 labourer	100/head	200
	Seeding cost	15 labourer	100/head	1500
2	Weeding			
	first	20 labourer	100/head	2000
	second	20 labourer	100/head	2000
3	Irrigation	4 numbers	300/irrigation	1200
		4 labourer	100/head	400
4	Plant protection	One spray	600/spray	600
		1 labour	100/head	100
5	Harvesting	20 labourer	100/head	2000
6	Threshing and winnowing	20 labourer	100/head	2000
7	Land rent for one season		500/ season	500
	Total A			**16100**
B	**Variable cost**			
1	**Wheat**			
i	Seed cost @ 100kg/ha	50 kg	12/kg	600
ii	Seed treatment by Bavistin	150 g @ 3g/kg of seed	450/kg	67
iii	Fertilizers			
	100:60:40 kg/ha N:P:K	N: 50 kg	12.13/kg	606
		P_2O_5: 30 kg	22.12/kg	664
		K_2O 20 kg	7.70/kg	154
		Total		**2091**
2	**Lentil**			
i	Seeding cost @ 40kg/ha	20 kg	50/kg	1000
ii	Seed treatment by Bavistin	60 g @ 3g/kg of seed	450/kg	27
iii	Fertilizers			
	30:50:30 kg/ha N:P:K	N: 15 kg	12.13/kg	182
		P_2O_5: 25 kg	22.12/kg	553
		K_2O 15 kg	7.70/kg	115
		Total		**1877**
	Total B			**3968**
	Grand total (A+B)- total cost Rs/ha		**92 labourer**	**20068**

Appendix –II d

	Cost of cultivation (Rs/ha) of wheat+ lentil (1:1) skip/alternate row of 20 cm (Rs/ha) of castor+lentil (1:3) 27%:73%			
SN	**Particulars**	**Input**	**Rate (Rs)**	**Cost (Rs)**
A	**Common or Fixed cost**			
1	Field preparation			
	Ploughing by tractor	4 hrs	400/hr	1600
	Harrowing and leveling	4 hrs	400/hr	1600
	cleaning of field	2 labourer	100/head	200
	Seeding cost	10 labourer	100/head	1000
2	Weeding			
	first	20 labourer	100/head	2000
	second	20 labourer	100/head	2000
3	Irrigation	2 numbers	300/irrigation	600
		2 labourer	100/head	200
4	Plant protection	One spray	600/spray	600
		1 labour	100/head	100
5	Harvesting	20 labourer	100/head	2000
6	Threshing and winnowing	20 labourer	100/head	2000
7	Land rent for one season		500/ season	500
	Total A			**14400**
B	**Variable cost**			
1	**castor**			
i	Seeding cost @ 20kg/ha	5.4 kg	65/kg	351
ii	Seed treatment by Bavistin	22 g @ 4g/kg of seed	450/kg	10
iii	Fertilizers			
	80:50:30 kg/ha N:P:K	N: 21.6 kg	12.13/kg	262
		P_2O_5: 13.50 kg	22.12/kg	299
		K_2O 8.10 kg	7.70/kg	62
		Total		**984**
2	**Lentil**			
i	Seeding cost @ 40kg/ha	29.20 kg	50/kg	1460
ii	Seed treatment by Bavistin	88 g @ 3g/kg of seed	450/kg	40
iii	Fertilizers			
	30:50:30 kg/ha N:P:K	N: 21.90 kg	12.13/kg	266
		P_2O_5: 36.5g	22.12/kg	807
		K_2O 21.9 kg	7.70/kg	169
		Total		**2742**
	Total B			**3726**
	Grand total (A+B)- total cost Rs/has		**95 labourer**	**18126**

Appendix –II e

Cost of cultivation (Rs/ha) (Rs/ha) of onion+coriander (3:1) 23%:77%				
S.N.	**Particulars**	**Input**	**Rate (Rs)**	**Cost (Rs)**
A	**Common or Fixed cost**			
1	Field preparation			
	Ploughing by tractor	4 hrs	400/hr	1600
	Harrowing and leveling	5 hrs	400/hr	2000
	cleaning of field	2 labourer	100/head	200
2	Weeding			
	first	20 labourer	100/head	2000
	second	20 labourer	100/head	2000
3	Irrigation	8 numbers	300/irrigation	2400
		8 labourer	100/head	800
4	Plant protection	One spray	600/spray	600
		1 labour	100/head	100
5	Harvesting	20 labourer	100/head	2000
6	Threshing and winnowing	20 labourer	100/head	2000
7	Land rent for one season		500/ season	500
	Total A			**16200**
B	**Variable cost**			
1	**Onion**			
i	Seed cost @ 8 kg/ha	6.16 kg	500/-	3080
ii	Seed treatment by Bavistin	20 g @ 3g/kg of seed	450/kg	10
iii	Nursery management			
	Field preparation			300
		10 labourer	100/head	1000
		irrigation + weeding		500
iii	Fertilizers			
	100:60:100 kg/ha N:P:K	N: 77 kg	12.13/kg	934
		P_2O5: 46.2 kg	22.12/kg	1022
		K_2O 77 kg	7.70/kg	593
iv	Transplanting	30 labourer	100/head	3000
	Interculture 1	20 labourer	100/head	2000
	Interculture 2	20 labourer	100/head	2000
		Total		**14439**
2	**Coriander**			
i	Seed cost @ 20kg/ha	4.6 kg	100/kg	460
ii	Seed treatment by Bavistin	15g @ 3g/kg of seed	450/kg	10
iii	Fertilizers			
	60:40:30 kg/ha N:P:K	N: 13.8 kg	12.13/kg	167
		P_2O_5:9.20 kg	22.12/kg	203
		K_2O 7 kg	7.70/kg	54
	Seeding cost	5 labourer	100/head	500
	Green leaf cutting	10 labourer	100/head	1000
		Total		**2394**
	Total B			**19333**
	Grand total (A+B)- total cost Rs/has		**176 labourer**	**33033**

Appendix –II f

Cost of cultivation (Rs/ha) (Rs/ha) of wheat+ Fenugreek (1:1) skip/alternate row of 20 cm 50%-50%				
S.N.	**Particulars**	**Input**	**Rate (Rs)**	**Cost (Rs)**
A	**Common or Fixed cost**			
1	Field preparation			
	Ploughing by tractor	4 hrs	400/hr	1600
	Harrowing and leveling	5 hrs	400/hr	2000
	cleaning of field	2 labourer	100/head	200
2	Weeding			
	Seeding cost	15labourer	100/head	1500
	First	20 labourer	100/head	2000
	second	20 labourer	100/head	2000
3	Irrigation	4 numbers	300/irrigation	1200
		4 labourer	100/head	400
4	Plant protection	One spray	600/spray	600
		1 labour	100/head	100
5	Harvesting	20 labourer	100/head	2000
6	Threshing and winnowing	20 labourer	100/head	2000
7	Land rent for one season		500/ season	500
	Total A			**16100**
B	**Variable cost**			
1	**Wheat**			
i	Seeding cost @ 100kg/ha	50 kg	12/kg	600
ii	Seed treatment by Bavistin	150 g @ 3g/kg of seed	450/kg	67
iii	Fertilizers			
	100:60:40 kg/ha N:P:K	N: 50 kg	12.13/kg	606
		$P2O_5$: 30 kg	22.12/kg	664
		K_2O 20 kg	7.70/kg	154
		Total		**2091**
2	**Fenugreek**			
i	Seeding cost @ 25 kg/ha	12.5 kg	70 /kg	875
ii	Seed treatment by Bavistin	38 g @ 3g/kg of seed	450/kg	20
iii	Fertilizers			
	40:40:30 kg/ha N:P:K	N: 20 kg	12.13/kg	243
		P_2O_5: 20 kg	22.12/kg	442
		K_2O 15 kg	7.70/kg	115
		Total		**1695**
	Total B			**3786**
	Grand total (A+B)- total cost Rs/ha		**102 labourer**	**19886**

Appendix –II g

Cost of cultivation (Rs/ha) of mustard+lentil (1:2)				
SN	**Particulars**	**Input**	**Rate (Rs)**	**Cost (Rs)**
A	**Common or Fixed cost**			
1	Field preparation			
	Ploughing by tractor	4 hrs	400/hr	1600
	Harrowing and leveling	4 hrs	400/hr	1600
	cleaning of field	2 labourer	100/head	200
	Seeding cost	15 labourer	100/head	1500
2	Weeding			
	first	20 labourer	100/head	2000
	second	20 labourer	100/head	2000
3	Irrigation	3 numbers	300/irrigation	900
		3 labourer	100/head	300
4	Plant protection	One spray	600/spray	600
		1 labour	100/head	100
5	Harvesting	20 labourer	100/head	2000
6	Threshing and winnowing	20 labourer	100/head	2000
7	Land rent for one season		500/ season	500
	Total A			**15300**
B	**Variable cost**			
1	**Mustard**			
i	Seeding cost @ 5kg/ha	1.75 kg	35/kg	62
ii	Seed treatment by Bavistin	6 g @3g/kg	450/kg	3
iii	Fertilizers			
	80:50:30 kg/ha N:P:K	N: 28 kg	12.13/kg	340
		P_2O_5: 17.5 kg	22.12/kg	387
		K_2O 10.5 kg	7.70/kg	81
		Total		**873**
2	**Lentil**			
i	Seeding cost @ 40kg/ha	26 kg	50/kg	1300
ii	Seed treatment by Bavistin	78 g @ 3g/kg of seed	450/kg	35
iii	Fertilizers			
	30:50:30 kg/ha N:P:K	N: 19.5 kg	12.13/kg	236
		P_2O_5: 32.5 kg	22.12/kg	719
		K_2O 19.5 kg	7.70/kg	150
		Total		**2440**
	Total B			**3313**
	Grand total (A+B)- total cost Rs/ha		**101 labourer**	**17113**

Appendix – II h

Cost of Cultivation (Rs/ha) of sunflower+lentil (1:3) 23%:77%				
S.N.	**Particulars**	**Input**	**Rate (Rs)**	**Cost (Rs)**
A	**Common or Fixed cost**			
1	Field preparation			
	Ploughing by tractor	4 hrs	400/hr	1600
	Harrowing and leveling	4 hrs	400/hr	1600
	cleaning of field	2 labourer	100/head	200
	Seeding cost	10 labourer	100/head	1000
2	Weeding			
	first	20 labourer	100/head	2000
	second	20 labourer	100/head	2000
3	Irrigation	3 numbers	300/irrigation	900
		3 labourer	100/head	300
4	Plant protection	One spray	600/spray	600
		1 labour	100/head	100
5	Harvesting	20 labourer	100/head	2000
6	Threshing and winnowing	20 labourer	100/head	2000
7	Land rent for one season		500/ season	500
	Total A			**14800**
B	**Variable cost**			
1	**Sunflower**			
i	Seeding cost @ 8kg/ha	1.84 kg	380/kg	700
ii	Seed treatment by Bavistin	5.5 g @ 4g/kg of seed	450/kg	3
iii	Fertilizers			
	80:80:60 kg/ha N:P:K	N: 18.4 kg	12.13/kg	223
		P_2O_5: 18.40 kg	22.12/kg	407
		K_2O 13.80 kg	7.70/kg	106
		Total		**1439**
2	**Lentil**			
i	Seeding cost @ 40kg/ha	30.80 kg	50/kg	1540
ii	Seed treatment by Bavistin	92.4 g @ 3g/kg of seed	450/kg	42
iii	Fertilizers			
	30:50:30 kg/ha N:P:K	N: 23.10 kg	12.13/kg	280
		P_2O_5: 38.5g	22.12/kg	852
		K_2O 23.10 kg	7.70/kg	178
		Total		**2892**
	Total B			**4331**
	Grand total (A+B)- total cost Rs/has		**96 labourer**	**19131**

Appendix – II i

Prices of test crops

S.N.	Crops	Prices of crops (Rs q^{-1})
1.	Rice	
	MTU 1010	1200
	Mahamaya	1000
	Indira sona	1000
2.	Wheat	1500
3.	Castor	3300
4.	Lentil	3600
5.	Mustard	2500
6.	Sunflower	2500
7.	Fenugreek	3500
8.	Onion	700
9.	Coriander	3000